# THE MODEL T

*Robert Casey*

# THE MODEL T
## A Centennial History

Johns Hopkins University Press
*Baltimore*

© 2008 The Henry Ford
All rights reserved. Published 2008
Printed in the United States of America on acid-free paper

Johns Hopkins Paperback edition, 2016
9 8 7 6 5 4 3 2 1

Johns Hopkins University Press
2715 North Charles Street
Baltimore, Maryland 21218-4363
www.press.jhu.edu

*The Library of Congress has catalogued the hardcover edition of this book as follows:*
Casey, Robert, 1946–
   The model T: a centennial history / Robert Casey.
      p. cm.
   Includes bibliographical references and index.
   ISBN-13: 978-0-8018-8850-2 (hardcover: alk. paper)
   ISBN-10: 0-8018-8850-6 (hardcover: alk. paper)
   1. Ford Model T automobile—History. I. Title.
   TL215.F75C37 2008
   629.222'2—dc22        2007049545

A catalog record for this book is available from the British Library.

ISBN-13: 978-1-4214-2117-9
ISBN-10: 1-4214-2117-8

All images courtesy of the collections of The Henry Ford, except as follows:

Page 4, courtesy of DaimlerChrysler.
Page 26, *bottom,* courtesy of Trent Boggess.
Page 119, *top* and *bottom,* respectively, courtesy of the Model T Ford Club of
   America and the Model T Ford Club International, Inc.
Page 121, courtesy of Primedia.

*Special discounts are available for bulk purchases of this book. For more information, please contact Special Sales at 410-516-6936 or specialsales@press.jhu.edu.*

Johns Hopkins University Press uses environmentally friendly book materials, including recycled text paper that is composed of at least 30 percent post-consumer waste, whenever possible. All of our book papers are acid-free, and our jackets and covers are printed on paper with recycled content.

*To the memory of Ford R. Bryan, who taught me that you are never too old to write your first book*

I will build a motor car for the great multitude. It will be large enough for the family but small enough for the individual to run and care for. It will be constructed of the best materials, by the best men to be hired, after the simplest designs that modern engineering can devise. But it will be so low in price that no man making a good salary will be unable to own one—and enjoy with his family the blessings of hours of pleasure in God's great open spaces.

Henry Ford, *My Life and Work*

**CONTENTS**

Foreword, *by William Clay Ford Jr.
  and Patricia E. Mooradian* xi
Acknowledgments xiii
Introduction xv

*1* **AUTOMOBILITY IN 1908** 1
*2* **CREATING THE MODEL T** 13
*3* **MANUFACTURING THE MODEL T** 32
*4* **SELLING THE MODEL T** 74
*5* **OWNING & DRIVING THE MODEL T** 96
*6* **THE MEANING OF THE MODEL T** 111

Notes 127
Further Reading 139
Index 143

*Color illustrations follow page 78.*

**FOREWORD**

Henry Ford had a simple idea. He wanted to build a car for the common man, a car "for the great multitude." Within a few short years, this simple idea had changed the world.

Henry Ford was born in 1863 into a world in which the fastest trains could barely achieve sixty miles per hour and most Americans lived on farms or in small towns and rarely traveled more than twenty-five miles from home during their lifetimes. By the time of Ford's death in 1947, airplanes were on the verge of exceeding the speed of sound, most Americans lived in cities, and families were dispersed across the continent.

The Model Ts that flowed out of Highland Park transformed the very nature of American society. They changed the way we live, work, and play. Henry Ford's "simple idea" created 20th-century America.

But Henry Ford did something more. A change-maker himself, he collected objects and stories of other innovators. In 1929 he founded the Edison Institute (now The Henry Ford) to "convey the inspiration of American genius to our young." Today, we use our sometimes rare, sometimes ordinary artifacts from America's past to tell stories of American ingenuity, innovation, and resourcefulness and to inspire the next generation of change-makers and innovators.

William Clay Ford Jr., Executive Chairman,
　Ford Motor Company
Patricia E. Mooradian, President, The Henry Ford

## ACKNOWLEDGMENTS

Like every author, I owe a great debt to a great many people. At the top of the list are my curatorial colleagues Bill Pretzer, Marc Greuther, Jeanine Miller, Dorothy Ebersole, Nancy Bryk, and Donna Braden. Their insights and encouragement were invaluable. They also picked up the slack while I was occupied with my research and writing. I hope I can return the favor in the future. John Metz added his encouragement and generously juggled his schedule to allow me to finish the manuscript.

In any bureaucracy a project needs a champion, and this project's champion was Judy Endelman, director of the Benson Ford Research Center. Her abiding support was absolutely essential. The staff of the Research Center was unfailingly helpful, especially Terry Hoover, Jim Orr, Kathy Steiner, Linda Skolarus, and Peter Kalinski. Cynthia Miller has an encyclopedic knowledge of the Research Center's massive photographic holdings. This book would not have been possible without her.

David Liepelt and Ken Kennedy patiently taught me how to drive a Model T and shared their extensive knowledge of the car's inner workings. Trent Boggess of Plymouth State University in New Hampshire has spent untold hours exploring the Research Center's Ford Motor Company holdings. He also owns a Model N and a Model T Ford. His rare combination of academic and practical knowledge provided many invaluable insights.

Most important of all was my wife, Julia. She supported me through the career change, job changes, and relocations that turned out to be prerequisites for this project. She put up with the late nights and weekends endlessly devoted to "the book." Henry Ford called his wife Clara "the Believer" because of the confidence she exhibited in him. I know what he meant.

## INTRODUCTION

You see them in parades on Memorial Day or the Fourth of July, often driven by men in derby hats or women in flapper-style dresses. Their engines chug, their horns go *ooogah*. Small children call them funny old cars, and parents remark that their grandparents (or great-grandparents) learned to drive in one just like that. Someone says, "They only came in black, you know," and others nod sagely.[1] Then they are gone. When the parade is over, spectators climb into their own automobiles and return to their suburban homes, perhaps stopping at a shopping mall on the way. And yet beyond some vague gratitude for "how far we have come," virtually no one makes any connection between the quaint, ungainly, chugging black Model T and the daily lifestyle everyone takes for granted.

Such is the fate of earthshaking technological devices after their shaking days are done. Make no mistake about it: Henry Ford's Model T was an earthshaking technological device. It filled deep, abiding desires that most people barely knew they had—desire for rapid, unfettered mobility; for control of something powerful; for ownership of something valuable, modern, and complex. Henry Ford felt in his bones the public's latent attraction for automobiles and believed that the key to tapping that attraction was an automobile both highly capable and inexpensive. So perfectly did his product align with the times that its sales made him the richest man in America. Yet before the Model T reached the end of its long production run, Ford had doubts about the consequences of what he had wrought. We share those doubts today and add many of our own. But we also share the deep, abiding desires that motivated Ford's customers in the first place.

The Model T cannot be separated from the man who created it. It was an expression of Henry Ford's ideas, sensibilities, and personality just as surely as a symphony or a sculpture or a novel expresses the ideas, sensibilities, and personality of the artist who creates them. While the car itself was seducing millions of buyers, the challenge of meeting the demand was seducing Ford and his engineers. In the end they created a production system that was as consequential as the car itself and that also expressed Ford's ideas, sensibilities, and personality.

Understanding the Model T story—how the car was designed, produced, and sold; why it died yet still lives—can, even after one hundred years, help us make sense of our world today. It is a great story. So open the door and climb aboard. Let's go for a ride.

# THE MODEL T

# *1* AUTOMOBILITY IN 1908

On October 1, 1908, Ford Motor Company introduced one of the most famous and influential products in the history of American business. Over its nineteen-year life cycle, the Model T made the company and its founder famous, wealthy, and powerful and altered American society forever. Before exploring the story of the car itself, let's take a look at the state of automobile technology at the moment that this revolutionary product was born.

Ford's latest design entered an American automobile industry only a dozen years old. The industry began in Springfield, Massachusetts, in 1896, when Charles and Frank Duryea became the first Americans to build a series of automobiles for sale.[1] The Duryea Motor Wagon Company made and sold thirteen cars in 1896. By 1898 the company dissolved and

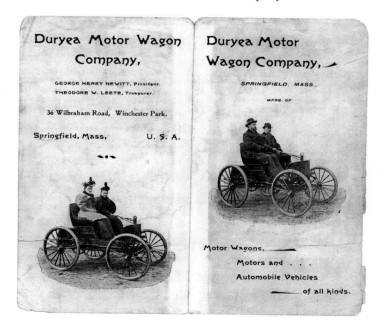

An 1896 brochure for America's first automobile company.

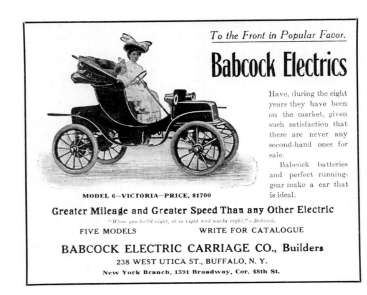

1908 Babcock Electric ad. Electric cars could no longer compete head-to-head with gasoline cars, so makers began to target women, who appreciated that electrics had no transmission to shift and needed no crank for starting.

the brothers went their separate ways, but other companies, such as the Winton Motor Carriage Company (1897) and the Haynes-Apperson Company (1898), entered the business. The industry grew steadily, if not spectacularly, selling some 63,500 cars by 1908.[2]

The March 1, 1908, issue of *Cycle and Automobile Trade Journal* contains a good snapshot of the state of the American auto industry and market. The magazine's "Eighth Annual Review of Complete Motor Cars" claimed to cover "all 1908 Models of all American Motor Car Manufacturers." It listed 166 makers of "pleasure cars"—that is, vehicles not designed for commercial use. Several trends emerge from an analysis of this listing.[3]

One is the utter domination of the gasoline-burning internal combustion engine. There were 151 companies making gasoline-powered cars, compared to 11 producing electric cars and only 4 manufacturing steam cars. Closer inspection reveals just how marginalized steam and electricity were. That electric cars had the most modern power source was veiled by their appearance, as they continued to look like horseless carriages, little different in appearance from the original gasoline cars of Duryea, Winton, and Haynes-Apperson. Electric cars were also expensive. None cost less than $1,150, and half the available models cost $2,000 or more.

Steam cars had on the whole adopted the look of their internal-combustion-powered brethren, but they were even more expensive than the electrics. Of the twelve different models of-

**THE INCOMPARABLE WHITE**

**THE CAR FOR SERVICE**

**Exclusive Features of the White Limousine**

The exclusive White quality of absolute noiselessness of operation is of particular advantage in a limousine because, in a car with a closed body, any noise made by the mechanism is even more noticeable and annoying than in an open vehicle.

Another exclusive White quality—namely—genuine flexibility of control, permits of the machine being guided safely and speedily through the crowded city streets. The speed of the White may be accommodated to the exigencies of street traffic without any changing of gears, jerky starts or the embarrassing and sometimes dangerous "stalling" of the engine.

As regards graceful lines and luxuriousness of equipment and finish, the White limousine must be seen to be appreciated.

Write for Catalog and the address of the nearest branch or agency.

**THE WHITE COMPANY**
CLEVELAND, OHIO.

New York City, Broadway at 62d Street
San Francisco, 1430 Market Street
Philadelphia, 629-33 North Broad Street
Boston, 320 Newbury Street
Chicago, 240 Michigan Avenue
Cleveland, 407 Rockwell Avenue

1908 White steamer ad. Note that the company never actually mentions that this car was powered by steam.

fered, only one, an $850 Stanley runabout, was priced less than $1,000, and only two other models were under $2,000.

The listings for gasoline cars reveal a market both diverse and unsettled. Gasoline vehicles came in three broad groups. Most numerous were the so-called Mercedes-type cars (sometimes referred to as "French-type"), inspired by Daimler's 1901 Mercedes. That German car diverged completely with horseless carriage design. Engineers mounted its 35-horsepower, 4-cylinder engine just behind the front axle rather than under the seat. This allowed them to lower the frame to only eight or nine inches off the ground and to seat the passengers behind the engine. The low center of gravity greatly improved the car's handling, so that the Mercedes could cruise at 50 miles per hour over good roads. American automakers recognized this design's advantages, but America lacked something basic: good roads.[4]

3 / Automobility in 1908

1901 Mercedes, built by Daimler. With a front-mounted 4-cylinder engine, lowered chassis, and seating for four, it set the pattern for future gasoline-powered cars.

The Mercedes was built for French and German roads, which were the best in the world. American roads were among the worst in the world. The second census of American roads, taken in 1909, classified only 8.66 percent of U.S. highways as "surfaced," which usually meant covered with gravel. The remainder were simply dirt: dusty in dry weather, muddy morasses in the rain, and creased with frozen ruts in the winter. To accommodate these abysmal thoroughfares, American engineers modified the Mercedes design in important ways. They raised the chassis to increase ground clearance, used thicker steel for frames and axles, made heavier springs, and favored larger-bore, shorter-stroke engines that yielded more power. By sacrificing speed for durability, they successfully adapted the European design to American conditions. Page after page of the *Cycle and Automobile Trade Journal* "review" depicts large four-passenger cars clearly patterned after the Mercedes. Like the German car, the American vehicles were also expensive, most costing between $2,000 and $7,500.[5]

At the other end of the car market was a peculiarly American automobile known as the high-wheeler, or the "western buggy type of automobile." These cars were even more primitive than the earlier horseless carriages. A typical high-wheeler featured a buggy frame, wooden wheels at least 36 inches in diameter, solid rubber tires, a rear-mounted engine (often air-cooled), tiller steering, and a sale price between $250 and $950. Most were made and used in the Midwest, hence the "western buggy" appellation. Salesmen touted their high ground clearance, light weight, and puncture-proof solid rubber tires as great advantages on dreadful rural midwestern roads. They enjoyed a brief heyday between 1905 and 1910. *Cycle and Automobile Trade Journal* listed twenty-three companies making thirty-one mod-

*(top)* In 1910 a Model T creates a dust cloud on one of the dirt paths Americans called roads.

*(bottom)* In the rain, American dirt roads turned into American mud roads. This Model T slogs through the muck in the 1909 New York-to-Seattle cross-country race.

(top) The 1908 Thomas, an American interpretation of the Mercedes-style car.

(bottom) Perhaps the best-known of the high-wheelers was the Holsman. This 1908 ad clearly illustrates why these cars were also known as "western buggy" types.

Looking like a miniaturized Mercedes-type car, this 1908 Maxwell was typical of the two-passenger runabouts.

els of high-wheelers. Unfortunately, high-wheelers tended to shake and vibrate fiercely, loosening nuts and bolts and breaking frames and suspensions. They also tended to be underpowered, ultimately disappointing their owners.[6]

In between the Mercedes-type cars and the high-wheelers were a group of cars known as runabouts. Most looked like smaller versions of the Mercedes-type cars, because many runabouts had their engines up front. Those that didn't often used false hoods to make their antiquated designs seem modern. With engines of one or two cylinders and prices between $600 and $1,000, they were far more rugged and capable than the high-wheelers but were less affordable to farmers. *Cycle and Automobile Trade Journal* listed twenty-seven models of runabouts made by twenty different companies. The runabout's primary disadvantage was that it had only two or (occasionally) three seats, thus limiting its usefulness to families.[7]

So even twelve years after the founding of the American automobile industry, the automotive landscape was muddled. No particular size or price range clearly dominated, and even though gasoline reigned as the preferred power source, steam and electricity still had their adherents. But the situation re-

Artist's rendition of how a farmer was likely to perceive the joy-riding automobilists from the city. As cars became more affordable, farmers' opinions changed.

flected more than just uncertainty about the right mix of size, power, and features. It also reflected uncertainty about what automobiles were for. One historian described the situation this way: "At stake were not only the forms motor-vehicle technology would take, but also the social ends it would serve. How, where, and with what effects, should people use the new machines?"[8]

Some Americans answered that question with a resounding "No how, no where, no way!" Opposition to the automobile came from a variety of places. Those with a stake in the horse-based economy—breeders, livery stable owners, teamsters, farriers—were natural enemies of the new technology. In cities, residents sometimes reacted violently to motorists who sped through lower-income neighborhoods where the streets were as much playground and meeting place as thoroughfare. The most widespread opposition came from farmers, who resented the intrusion of drivers, who usually came from urban centers. As historian James Flink put it, "Speeding automobile tourists constituted a danger both to stock and to horsedrawn traffic and raised clouds of dust that damaged crops and settled on farmhouses, barns, and washes hung out to dry." Like the residents of urban neighborhoods, farmers believed roads existed for local social and economic purposes, not the convenience of joy-riding outsiders. Other critics held the car responsible for larger problems. Future president Woodrow Wilson feared that the obvious gap between wealthy motorists and those who could not afford cars was breeding socialism.[9]

But opposition to the "auto" declined as actual car ownership, along with the hope of future car ownership, spread throughout the country. The change was especially apparent among farmers. When the cheap high-wheeler put automobility within reach, many farmers revised their anti-auto attitudes. By 1908 the leader of New Jersey's chapter of the largest farmers' organization, the National Grange, was saying, "We farmers are not opposed to the motorcar... it will be an important feature in making farm life more attractive. When the motorcar becomes cheaper in price through more general use the farmer will be the first to adopt it for business and pleasure. I stand ready to promise the motorists that the grangers of this state will cooperate with them in bettering not only the condition of the roads but the lot of the motorists, who it appears, have been legislated against too severely."[10]

Advocates for the automobile tended to fall into one of two groups: "horse-minded" people, who saw cars as a direct replacement for horses, and "transportation-minded" people, who evaluated cars in the context of all other forms of transportation.[11]

Horse-minded auto advocates credited the motor vehicle with several advantages over Dobbin: it didn't eat when it wasn't running; it didn't defecate and urinate all over city streets; it could be parked at home rather than boarded at a livery stable; it wasn't skittish and didn't run away at inopportune times. Thus the motor vehicle should logically supplant horses in cities. A few users who needed transportation at a moment's notice, such as physicians, found that cars could indeed be direct replacements for horses. More often, however, the auto's theoretical advantages didn't play out in practice.[12]

For instance, efforts to sell clean, quiet electric cars as city transportation came to naught. An attempt to monopolize the taxi business in major cities by replacing horse-drawn cabs with electric ones (the Lead Cab Trust, as it was called) foundered due to financial mismanagement and the poor quality of the vehicles. Attempts to sell individual owners on the merits of motor vehicles for urban transportation fared little better. As cities developed, they of necessity adapted themselves to the limits of the available transportation means. Thus, in American cities of the early twentieth century, trolleys, horse-drawn cabs, even bicycles and "shank's mare" got people everywhere they wanted to go because all the places they wanted to go were located within reach of those transportation modes. The expen-

Trolleys and horse-drawn vehicles outnumber cars in 1911 Chicago. Urban areas were not initially fertile markets for automakers.

sive private automobile offered no obvious advantages over the established conveyances. Similarly, motor trucks, whether powered by steam, gasoline, or electricity, successfully challenged horse-drawn commercial vehicles only after World War I.[13]

Transportation-minded motor vehicle advocates thought more broadly. They recognized the potential of automobiles to do things other transportation modes could not do. Often these were things car owners *hadn't known they wanted to do.* For instance, the sheer joy of controlling a powerful, rapidly moving machine proved irresistible to many. As one advocate explained, "It is in the running of the car, the handling of it, and its obedience to one's will, that the keenest enjoyment of automobiling is found." Automobiles also gave owners a new outlet for the individuality that has long been a core value of American society. A contributor to *Harper's Weekly* wrote that his automobile offered him "the feeling of independence—the freedom from timetables, from fixed and inflexible routes, from the proximity of other human beings than one's chosen companions; the ability to go where and when one wills, to linger and stop where the country is beautiful and the way pleasant, or to rush through unattractive surroundings, to select the best places to eat and sleep; and the satisfaction that comes from a knowledge that one need ask favors or accommodation from no one nor trespass on anybody's property or privacy."[14]

These writers were speaking about the automobile less as

This 1910 Ford brochure emphasizes driving as recreation.

a means of transportation than as a means of recreation. Put another way, they were speaking not of auto travel that *needs to be* done but of travel that *can be* done. This role as recreational device proved crucial to the automobile's development in the first decade of the twentieth century. Affluent city dwellers who didn't need cars to get around on a daily basis used them to take rural excursions. As one car owner put it, "To possess a car is to become possessed of a desire to go far afield. The limits of the city become narrow, contracted, cramped, cagelike. The desire, so to speak, to spread its wings is in the nature of the motorcar, if things inanimate may be said to be moved by desire." It was, of course, just such joy-riding city dwellers that initially set farmers against automobiles. But farmers proved as susceptible to the lure of recreational travel as city folk. In 1907 a newly minted driver noted that he once had been "perfectly content to have my horizon bounded by the ten- or twelve-mile circle that my horses could go in a day and get back in good shape. I now make exploring trips of five times the distance. I can call on friends living thirty miles away who have been asking me to come for twenty years." It is no surprise, then, that *Cycle and Automobile Trade Journal* divided its listing of new vehicles into "commercial" and "pleasure," because private automobiles were first and foremost a source of pleasure.[15]

The rise of recreational driving over rural roads was the decisive factor in the ultimate triumph of the gasoline-powered auto. Electric cars that depended on batteries were obviously unsuited to traveling over rough, hilly country roads with no power lines and no recharging stations. Steam cars appeared to stack up well with gasoline cars in the early part of the century. They were fast and virtually impossible to stall, even on steep hills or in bottomless mud. But early steam cars lacked condensers, so they needed to be replenished with clean water at least as often as they needed fuel. Gasoline cars needed only the fuel, which was readily available even in country stores. The Mercedes-style gasoline cars that replaced the gasoline horseless carriages matched most of the steamers' strengths while avoiding their weaknesses. So it was that by 1908 the superiority of the Mercedes-style car as a recreational vehicle had driven both steam and electricity to the margins of the automotive universe.[16]

By 1908 the automobile market was dominated by cars catering to the preferences of well-to-do, transportation-minded urban dwellers. But that market segment was nearing saturation. Meanwhile, desire for automobiles was growing among the larger group of less affluent horse-minded farmers and small-town residents. Rural people wanted to use automobiles the way they used horses—as basic transportation. But they could not afford large, Mercedes-style cars, and they were painfully aware of the limitations of high-wheelers and runabouts. What rural dwellers wanted was something in between, a compromise vehicle that one publication called the "lightweight automobile." In 1904, *Outing* magazine speculated that "it is not impossible that a system of construction may be perfected to a point where it will give a four-passenger car of reasonable efficiency and moderate speed at a cost of not very much over one thousand dollars." To acquire the benefits of automobility, many potential buyers would be "content to dispense with some of the more sensational features of motoring so long as they can travel safely and surely."[17]

Four years later, that "four-passenger car of reasonable efficiency and moderate speed" arrived in the form of the Model T. Priced well under $1,000, it was a classic example of the right product at the right time.

# 2 CREATING THE MODEL T

Wearing the proper mustache of the day, Henry Ford poses in 1893 with two of his employees at the Detroit Edison Illuminating Company. At the far right is George Cato, who would help Ford build his first horseless carriage in 1896.

While the Duryea Motor Wagon Company was building the first American production automobiles, the chief engineer of the Detroit Edison Electric Illuminating Company was working in his spare time to make a horseless carriage that would simply run. Henry Ford, working in a shed behind his house in Detroit, was one of many young men across the country experimenting with the new transportation technology. Ford completed his gasoline carriage in 1896, followed by a second one in 1898. The next year he left Detroit Edison to start the Detroit Automobile Company. The DAC failed within a year, however, so Ford organized the Henry Ford Company. In 1902 disagreements with his financial backers led Ford to leave that

Ford poses in October 1896 with the Quadricycle, his first working automobile. Note the chain drive, tiller steering, and electric bell as a warning device. The engine is enclosed underneath the seat.

firm. Despite these business setbacks, Ford kept his engineering reputation alive with successful racing cars in 1901 and 1902. In the latter year, Ford connected with Alexander Malcomson, a prosperous coal dealer with a penchant for taking risks. Malcomson brought new money and the ability to attract new investors, while Ford brought his evolving ideas about automobiles. In 1903 they formed the Ford Motor Company.[1]

The new company's early vehicles, the 2-cylinder Models A, C, and F followed typical horseless carriage practice: they mounted their engines underneath the seats, driving the rear wheels through a chain. But a sharp disagreement over future product strategy developed within the company. Malcomson and four fellow stockholders, favoring the construction of large, expensive cars, pushed the development of a 4-cylinder Model B (priced at $2,000) and a 6-cylinder Model K ($2,500). Ford, backed by six other stockholders, championed cheaper cars, especially the small, 4-cylinder Model N, which he intended to sell for $500. Relations between the two factions deteriorated, and Henry Ford concluded that Malcomson had to go.

Ford and the small car faction devised a clever strategy to force the coal dealer out. In November 1905 they formed a new venture, the Ford Manufacturing Company, that did not include Malcomson and his large car cohorts. The Manufacturing Company would make parts for the Model N and sell them to the Motor Company. But the Manufacturing Company set the prices for its parts high enough to eat up most of the Model N's profits. Put another way, the Model N's profits

Ford Motor Company's first product, the 1903 Model A, as depicted in one of the company's first pieces of sales literature.

Despite its Mercedes-style front hood, the 1904 Model C's engine is under the seat, as with other horseless carriages.

Ford literature tried to present the 1906 Model F as a modern car. But take away the rear seat and the false front hood, and the car is not fundamentally different from the 1903 Model A.

A 1905 brochure illustrates Ford's Model B, the company's first 4-cylinder, Mercedes-style car.

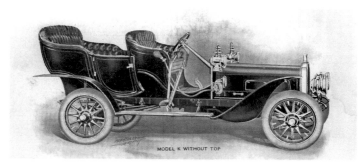

Alexander Malcomson advocated the production of the big 6-cylinder Ford Model K. It looked impressive in this 1906 company publication, but it was a consistent money-loser.

passed from the Motor Company through the Manufacturing Company to Ford and his allies. That left the Motor Company to survive on whatever income the slow-selling Model K generated. The revenues were so meager that Malcomsom and his supporters would see no dividends at all. Frustrated and outmaneuvered, Malcomson capitulated in July 1906, selling his stock to Ford. In the next few months, the coal dealer's supporters also sold out. In May 1907 the Ford Manufacturing Company, having served its purpose, sold all of its assets to the Ford Motor Company and ceased to exist. Malcomson, who had believed in Henry Ford when Ford was a two-time loser in business, was the first of a long line of people to learn a hard lesson: don't get between Henry Ford and his vision.[2]

What was the vision that Henry Ford pursued? Large-scale manufacturing was part of it. In his autobiography Ford wrote that at age nineteen he investigated the possibility of manufacturing large numbers of inexpensive watches. He ultimately rejected the idea, but the challenges of large-scale production continued to interest him. "Even then," Ford noted, "I wanted to make something in quantity." In 1903 he expounded to the Ford Motor Company stockholder John W. Anderson on the importance of uniformity, saying that cars should "come through the factory just alike; just as one pin is like another pin when it comes from a pin factory." Over the next three years the vision

Specifications, Model N

A 1906 Ford brochure depicts a Model N, the car that precipitated the conflict between Henry Ford and Alexander Malcomson.

A page from a 1906 Maxwell catalog advertising the Model L.

crystallized. In a 1906 letter to *The Automobile* magazine, Ford wrote that the "greatest need today is a light, low priced car with an up-to-date engine with ample horsepower, and built of the very best material.... It must be powerful enough for American roads and capable of carrying its passengers anywhere that a horse-drawn vehicle will go without the driver being afraid of ruining his car."[3]

That description fits the Model T, but it also fits the Model T's predecessor, the Model N. The car that sparked the terminal split between Henry Ford and Alexander Malcomson was an immediate hit with auto observers and buyers. *Cycle and Automobile Trade Journal* called it "distinctly the most important mechanical traction event of 1906." The Model N offered amazing value: a two passenger runabout with four cylinders, 15 horsepower, and a shaft drive capable of 45 miles per hour and priced at only $500. A comparison with other popular low-priced two-passenger cars demonstrates what a bargain the Model N was.[4]

Maxwell charged $780 for its Model L, a 2-cylinder, 10-horsepower car with a shaft drive. Buick wanted $1,250 for the 2-cylinder, 22-horsepower Model G with an older-style chain drive. The $500 Brush, advertised as "Everyman's Car,"

This 1906 Buick catalog highlights the Model G's tilting steering wheel that made entry and exit more convenient.

The $500, 1907 Brush competed on price with Ford's Model N. But readers of this Brush brochure also learned that the car had a wooden frame, a 1-cylinder engine, and solid rubber rather than pneumatic tires.

was a 1-cylinder, 6-horsepower, chain drive vehicle featuring a wooden frame and wooden axles to keep costs low.[5] Even though Ford ultimately had to raise the Model N's price to $600, the car was a sales success. But the Model N, as well as its slightly upscale variations Models R and S, were all still runabouts, seating at most three people. They lacked the room and the horsepower that people with families wanted. Henry Ford believed he could do better, so he set about designing a new model.

In creating the Model T, Henry Ford used an approach he had been developing since childhood. While in elementary school he induced his friends to assist him in projects like damming a small stream to power a water wheel or building a rudimentary boiler and steam turbine. The friends did most of the work, but they followed Henry's directions. At least four men—Jim Bishop, George Cato, Ed "Spider" Huff, and David Bell—assisted Ford in building his first car (which he dubbed the Quadricycle) in 1896. Bell recalled that he "never saw Mr. Ford make anything. He was always doing the directing." Ford associate Oliver Barthel made similar comments about the building of Ford's first race car in 1901. One of Ford's greatest talents was his ability to conceive a vision, articulate that vision to others (who often had talents and skills Ford himself lacked), and imbue them with the desire to work very hard to achieve his vision.⁶

To create the Model N's successor, Ford assembled a small group of men who were well trained, mechanically adept, and young. C. Harold Wills, a toolmaker who studied chemistry and metallurgy at night, was just twenty-one when he joined Ford at the Detroit Automobile Company in 1899. Hungarian immigrant Joseph Galamb brought a degree from the Industrial Technology school in Budapest, an expressive style as a draftsman, and experience in a variety of German and American industrial plants. Wills hired him at age twenty-four in 1905. Henry Ford brought twenty-four-year-old Charles Sorensen aboard in 1905. The Danish-born Sorensen brought a wealth of experience as a foundry pattern maker. Ed "Spider" Huff may have had the least education of all the Model T's creators, but one of his contemporaries described him as a genius. He was only sixteen when he worked on Ford's Quadricycle in 1896, only twenty-eight when he designed the T's magneto.⁷

One Ford employee who had no part in the Model T's design but played a great part in its ultimate success was James Couzens. As Alexander Malcomson's clerk, he oversaw the coal dealer's interests in the new Ford Motor Company. Seizing the main chance, Couzens took over the business side of the operation, freeing Henry Ford to concentrate on manufacturing and engineering. Couzens sided with Ford in the battle over the Model N and ultimately became the company's second-largest stockholder. He left Ford in 1915 to enter politics, but his energy, organizational skill, and business acumen were essential in setting the company on the road to success.

The unsung hero in the Model T story might be an English metallurgical engineer named J. Kent Smith. Smith decisively demonstrated to Ford the advantages of vanadium alloy steel. Ford related that he first encountered vanadium steel when he picked up a part from a wrecked French racing car. Harold Wills told a more prosaic story of learning about the alloy at an engineering conference in 1905. Charles Sorensen said that information about vanadium steel developments in England was available in engineering journals by 1905. Whatever the truth, Henry Ford was intrigued by the possibilities of a steel touted as being lighter and stronger than standard carbon steel. Smith visited Ford's Detroit factory on Piquette Avenue in 1906 and discussed the new steel with Ford and Wills. Ford made a trip to Smith's laboratory in Canton, Ohio, where tests demonstrated the truth of the claims for vanadium steel. Ford acted quickly, adopting the new alloy for axles and gears in the

1907 Model N and Model K. The new steel would play a much bigger role in the coming Model T.[8]

Early in 1907 Henry Ford ordered that a room be built in the northeast corner of the Piquette Avenue Plant's third floor. Behind its padlocked door he installed Joe Galamb and his drafting table. Besides Galamb, only a handful of people had access to the room: Wills, machinist C. J. Smith, Ford and his fourteen-year-old son Edsel, and perhaps Sorensen. In that room they created the Model T.[9]

Development proceeded in a straightforward fashion. Henry brought his ideas and concepts. Galamb drew them, often on a blackboard. Ford and his colleagues critiqued the designs, making changes. At some point Ford ordered machine tools, like a lathe and a milling machine, moved into the room. Prototype parts were built, sometimes machined from stock by Smith, sometimes made from wood by pattern maker Sorensen, sometimes cast in metal. As Galamb noted, Ford "liked to see a model working first. He didn't like to go just by the blueprint. He never did. He always liked to have a sample made first." A Model N chassis provided the initial test bed for prototype parts, but by October 1907 two hand-built Model Ts were ready for testing.[10]

People typically describe the vehicle that emerged from the small experimental room as practical, utilitarian, or cheap. This characterization is true but incomplete, because it ignores the aesthetic dimension of the Model T's design. Anyone gazing upon the tall, squarish, ungainly black bulk of a Model T Ford might well ask, "What aesthetic dimension?" But contrary to Ford's own claim that "Edsel is the artist in our family. Art is something I know nothing about," several observers did speak of Henry Ford as an artist. As naturalist John Burroughs, one of Ford's companions on a series of well-publicized camping vacations, wrote in 1918, "No poet ever expressed himself through his work more completely than Mr. Ford has expressed himself through his car and his tractor engine—they typify him—not imposing, not complex, less expressive of power and mass than simplicity, adaptability, and universal service."[11] Biographers Allan Nevins and Frank Hill made the case more comprehensively: "The dreamer, the man of intuitive mind, is usually an artist; and many vagaries, many contradictions, many triumphs and failures, become comprehensible in Ford if we view him as a man of imaginative and artistic temperament. His detachment, his wry humor, his constant self-dramatization, his ability to see affairs in large terms, and above all his creative zest, all bespeak an artistic bent."[12]

Popular opinion assumes that engineering designs result from logical calculations and the rational discovery of the "best" way to solve a particular problem. Artistic designs, by contrast, result from non-rational forces—feelings, emotions, whatever looks right to the artist. Engineer-turned-historian Eugene Ferguson exposed the falsity of this dichotomy: "Design engineers have recourse to analytical calculations to assist them in making decisions, but the number of decisions that are based on intuition, a sense of fitness, and personal preference made in the course of working out a particular design is probably equal to the number of artists' decisions that engineers call arbitrary, whimsical, and undisciplined."[13] That "intuition" and "sense of fitness, and personal preference" is different for each engineer, as it is for each artist, and constitutes a personal engineering aesthetic. Each engineer knows what "looks right" subjectively, even if he or she can't always explain why it looks right. The designs that looked right to Henry Ford always featured light weight and simplicity.

Metallurgist John Wandersee noted that "Mr. Ford was always for a light car because it could cut rings around the big cars." Joseph Galamb told the story of Ford ordering an experimental Model T crankshaft that was a quarter inch smaller in diameter to save weight. The resulting crankshaft was too flexible for practical use, but as Galamb said, Ford "wanted to get things light every way he could." Ford often justified his preferences on practical grounds, as when he said, "The less complex an article, the easier it is to make, the cheaper it may be sold, and therefore the greater number may be sold." But he could also speak in purely aesthetic terms: "The most beautiful things in the world are those from which all excess weight has been eliminated."[14]

With these thoughts in mind, it is useful to return to Ford's *Automobile* letter from 1906 and see how the Model T stacked up against the specifications outlined: "a light, low priced car with an up-to-date engine with ample horsepower, and built of the best material...powerful enough for American roads and capable of carrying its passengers anywhere that a horse-drawn vehicle will go without the driver being afraid of ruining his car."

Unsurprisingly, the first design requirement Ford mentions is light weight. The finished Model T touring car weighed only 1,200 pounds. No other popular four-passenger car was even close. For instance, the Buick Model 10 Tourabout ($1,050) weighed 1,570 pounds, and the Overland Model 32 Toy Tonneau ($1,500) weighed 1,750 pounds. Even the flimsy high-

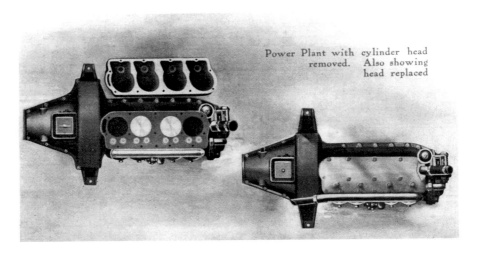

Power Plant with cylinder head removed. Also showing head replaced

Catalogs for the new Model T emphasized the detachable cylinder head. A copper-asbestos gasket sealed the joint between the head and the block.

wheelers exhibited little advantage over the Model T. The four-passenger version of the popular Holsman weighed 1,100 pounds and cost $740.[15]

Without question, the Model T's engine was "up-to-date" with "ample horsepower." The engine's most striking features were its one-piece cylinder block and its detachable cylinder head. These eventually became standard practice, but in 1909 they represented real advancements. Most contemporary cars, regardless of cost, used cylinders cast singly or in pairs and bolted to a separate crankcase. The resulting engines were large, heavy, and costly to manufacture. *Horseless Age* noted that among the cars exhibited at the 1909 Grand Central Palace show in New York, only 7 percent had engines cast in one piece, Ford's being the only American car to boast such an engine.[16] The detachable cylinder head was even more unusual. As late as 1911 a standard treatise on gasoline automobiles noted only two American cars with detachable heads: the Ford and the much more expensive Knox.[17] The Ford's removable head offered advantages for both producer and consumer. The separate block and head were simpler and cheaper to cast and machine than the typical one-piece units. They also made maintenance tasks such as grinding the valves easier to perform.

The Model T engine's 20 horsepower hardly seems "ample" today, but coupled with the 1,200-pound weight of the finished car, it made the Ford a lively performer. Typical American cars of the day weighed about 80 pounds for each horsepower. The Model T's 60 pounds per horsepower stacked up well against expensive cars like the Thomas Flyer that won the New York–to-Paris race in 1908. The Thomas weighed 64 pounds for each horsepower, but it cost $3,500.[18]

The Model T was indeed "built of the very best material."

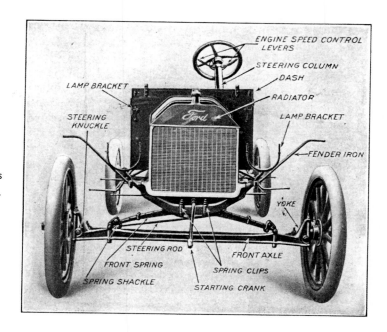

The front suspension of most cars of the day used a pair of leaf springs mounted perpendicular to the axle. Ford's single spring, mounted parallel to the axle, was both lighter and simpler. This illustration and several of the next few images come from Victor Pagé's *The Model T Ford Car*, one of the popular handbooks available to Model T drivers.

The vanadium steel referred to above played an important role, but so did the heat-treating techniques the company developed. Relying on his intuition about men's talents, Henry Ford selected John Wandersee, a man with no previous experience in metallurgy, to be trained as a metallurgist. Under Wills's supervision, Wandersee and August Degener developed ways of heat-treating vanadium and other steels to tailor their chemical composition and physical characteristics for specific uses. These efforts produced strong, tough parts no larger or heavier than they needed to be.[19]

As we noted in chapter 1, America had dreadful roads. So Ford's prescription that a car be "capable of carrying its passengers anywhere that a horse-drawn vehicle will go without the driver being afraid of ruining his car" was no small requirement. To deal with deep ruts, Ford gave the Model T plenty of ground clearance: ten inches between the road surface and the lowest point on the vehicle. Harder to solve was the problem of creating a chassis capable of absorbing the punishment meted out by those dreadful roads. Most carmakers resorted to building big, strong, rigid chassis. But that approach conflicted with Ford's desire to make his car inexpensive as well as with his personal aesthetic preference for light weight. Instead, the Ford team created a light, flexible chassis designed to absorb pounding without tearing itself apart. Drawing on experience gained with the Models B and N, Ford and his people based the new car's suspension and engine mounting on a series of triangles.

Images on pages 23 and 24 show the Model T's front sus-

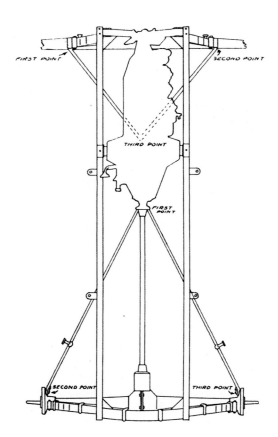

Axles and radius rods formed strong triangles at the front and rear of the Model T. *Fordowner* magazine frequently carried diagrams like this one from the July 1915 issue.

pension. The single leaf spring allowed the axle and wheels to move with the contours of the road. The axle and the two stiffeners known as radius rods formed a triangle that kept the axle ends from moving back and forth and kept the axle from twisting about its own axis. The Model N used a similar system, and Ford transferred it to the Model T with little significant change.

The rear axle, shown above and on page 25, used a similar arrangement. The major addition was the drive shaft that connected the transmission to the differential housing. A tubular housing known as a torque tube enclosed the driveshaft, and a universal joint connected the torque tube to the transmission. As with the front suspension, the triangle formed by the radius rods and the axle allowed the axle ends to move up and down but not back and forth and kept the axle from twisting. Versions of this arrangement appeared on the Models B and N with two parallel elliptical springs rather than the single transverse spring.

The final triangle involved the Ford motor itself. On most cars of the day, the motor mounted to the frame at four points,

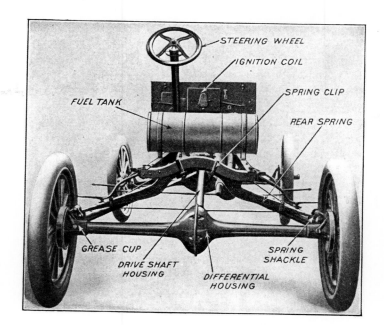

The Model T's rear suspension was similar to the front suspension. Once again, Ford's use of a single spring parallel to the axle was lighter and simpler than the typical industry practice of two springs perpendicular to the axle.

two on each side. Any twisting of the frame resulting from ruts or bumps was transmitted directly to the motor, sometimes breaking the mounts. As shown on page 26, *top,* Ford engineers attached their motor at only three points: two in the rear and one in the front. The front mount was not rigid but was a cylindrical bearing resting in a trunnion mounted on the frame's front cross-member. Thus, when road conditions caused the frame to twist, the twist was not transferred to the motor. This arrangement improved on the system developed for the Model N, which lacked the front bearing and trunnion and sometimes resulted in broken front cross-members.

These three sets of flexible mountings proved essential to the Model T's success and longevity. Like a reed bending in the wind, the Ford chassis twisted with the ruts, holes, and bumps of American roads, but did not break. At the bottom of page 26, we see the Ford chassis in action.

Henry Ford's twin lodestones of simplicity and light weight drove the development of two of the Model T's most idiosyncratic features, its transmission and its magneto.

The internal combustion engine has one inherent weakness as an automobile power plant. It delivers its maximum driving effort (what engineers call torque) at high speeds and its minimum driving effort at low speeds. But an automobile needs maximum torque when starting from rest (zero speed) and needs much less at higher speeds. So an internal combustion engine requires some arrangement of gears or pulleys to increase torque to the driving wheels at low speeds and de-

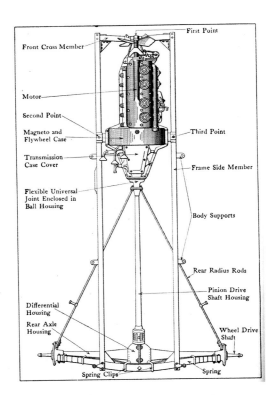

*(top)* This drawing illustrates the three-point engine mounting. The flexible bearing and trunnion are identified as the "First Point," while the "Second Point" and "Third Point" are where the engine was mounted rigidly to the frame.

*(bottom)* Modern Model T enthusiasts sometimes seek out the sort of roads for which their cars were designed. The car's suspension adjusts to the uneven terrain without unduly stressing the frame or the engine.

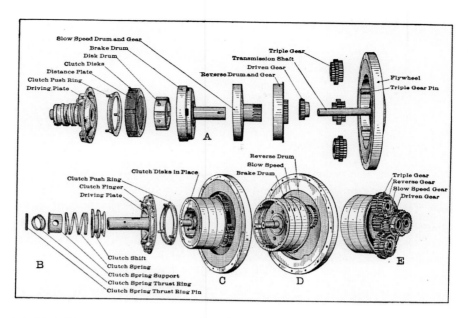

An exploded view of the Model T's planetary transmission.

crease torque to the wheels at high speeds. This set of gears or pulleys is called the transmission. Also needed is some device for temporarily disconnecting the engine from the transmission when the car stops or when the driver shifts the gears from one speed to another. That device is the clutch. Both the clutch and the transmission posed major design problems for early automobile builders.

The most popular form of transmission was the sliding gear type, in which a lever moved spinning gears from one position to another. Making these gear changes smoothly, quietly, and without damage to the gears themselves required practiced, coordinated movement of the lever, the clutch, and the engine throttle. An alternative was the planetary transmission, whose gears were always in mesh. Drivers changed speeds by means of brakes (usually called bands) that stopped or released shafts connected to the gears. Planetary transmissions were easy to shift but were generally not rugged enough for use on larger cars and at this time were limited to two forward speeds.[20] Henry Ford greatly favored planetary transmissions. All previous Ford Motor Company vehicles had them, and the lightweight new Model T would as well. Joseph Galamb observed that the planetary transmission was the first Model T component to be designed.[21]

Most early horseless carriages employed planetary transmissions, but their use declined as cars grew larger. At the two 1909 New York City auto shows, only 15 out of 170 models featured planetary transmissions.[22] But Ford knew his market. He designed his Model T for people with no driving experi-

27 / Creating the Model T

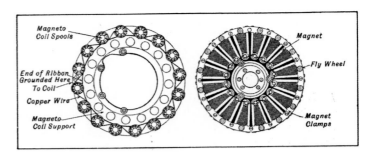

The essential elements of the Model T magneto. When the flywheel rotated, the magnetic fields generated by the V-shaped magnets passed through the stationary wire coil spools, generating electric current for the spark plugs.

ence whatsoever. For them the easy-to-learn planetary transmission was a wise choice. A 1909 letter to the editor of *Horseless Age* illustrates the complications of operating a sliding gear transmission. The correspondent consumes five full columns explaining how to shift gears noiselessly. He concludes by recommending that the driver have a "mental picture of just what transpires in the gear box every time he moves the lever or touches the clutch or moves the car."[23] Ford knew that the great mass of potential auto buyers had no interest in forming such "mental pictures." They simply wanted to drive their car, and the easy shifting planetary transmission allowed them to do that.

The most original element in the Model T's design was probably the flywheel magneto. It generated the electricity that ignited the gasoline/air mixture inside the engine's cylinders.

The magneto (and all electrical generators) works on a principle first documented by Michael Faraday in 1831. When a coil of wire passes through a magnetic field (or vice versa), an electric current is induced in the coil. Most automotive magnetos were discrete devices mounted under the hood of a car and driven off the engine with a belt or gears. Ford's solution was characteristically lighter and more compact. He built the magneto into the engine flywheel. The flywheel itself carried sixteen V-shaped permanent magnets. Mounted at the rear of the engine were sixteen stationary wire coils arranged in a circle. When the flywheel revolved, the magnetic fields from the magnets passed through the coils, generating the needed current.

This basic concept went back at least as far as December 1903, when Ford met Vincent G. Apple, an engineer with an idea for a flywheel magneto. Ford supplied Apple with a 2-cylinder Model A engine to try it out, but the two engineers never came to an agreement, so Ford moved on. In 1904 Charles Marie François Aufière received a French patent on a flywheel magneto closely resembling the final Model T device. His work may have influenced Ford's efforts. The job of designing the Model T magneto fell to Spider Huff, who worked as an outside consultant, not as an employee of the Ford Motor

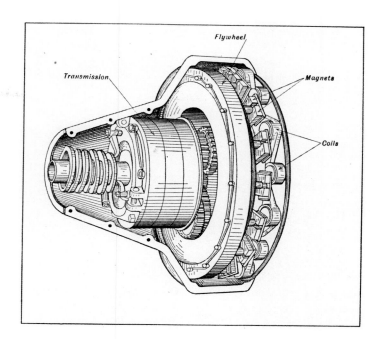

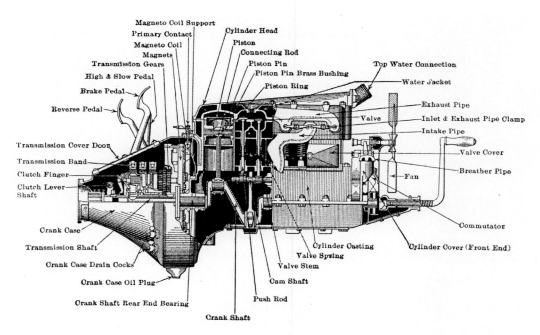

The transmission and magneto formed a compact assembly.

The transmission, magneto, and engine were mounted together as a unit, all lubricated by the same oil.

Company. As is often the case, a simple concept turned out to be complex to execute, and it seems to have taken Huff over a year to develop a reliable magneto. He applied for three different patents related to the flywheel magneto, on December 10, 1907, and March 2 and April 27, 1908.[24] The engine, magneto,

This 1909 Ford brochure touts left-hand drive as both safer and more convenient for the passenger.

and transmission were mounted together end to end, enclosed as a compact, lightweight, oil-tight, dust-tight package.

Curiously, no record survives explaining the decision to put the Model T's steering wheel on the left side of the car. All previous Fords had had right-hand steering. That was the standard arrangement for American cars, even though Americans drove on the right side of the road, as they do today. At the 1909 New York auto shows, only two other cars featured left-hand steering.[25] Perhaps Ford's choice is best explained by the company's own advertising. A 1909 Ford brochure illustrates the advantages of having the passenger, in this case a woman, enter and exit a Ford from the curb (right-hand) side, away from traffic and the muddy street. The text also explained that left-hand drive allowed the driver to judge more accurately the distance between the driver's car and one passing in the opposite direction and gave the driver a better view of oncoming or overtaking traffic when passing or turning left. Over the next few years, no doubt because of the Model T's vast popularity, left-hand steering became standard in the United States.

The first running prototype Model Ts came out of the Piquette Plant in October 1907. Development work continued for another year, even after serial production began. For instance, the *Advance Catalog* sent to dealers in March 1908 shows the transmission and brakes being operated by three foot pedals and two hand levers. But the car entered production with two foot pedals and two hand levers, and plans were already under way to convert to the three-pedal, one-lever system remembered by the vast majority of Model T owners. The first 2,500 production cars used a gear-driven pump to circulate the cooling water. Ford's preference for simplicity led him to delete the pump in all the succeeding cars in favor of the so-called thermo-syphon

The Model T prototype after its September 1908 road trip. The car is parked outside Ford's Piquette Avenue Plant in Detroit.

system, which utilized natural circulation caused by the water in the engine being hotter than the water in the radiator. Ford also learned that he could carry his obsession with light weight too far. The steel used for the frames of the first 2,500 Model Ts was the same thickness as the steel used for Model N frames. This proved to be too light, and all 2,500 had to have reinforcing plates riveted inside the frame rails. Ford used thicker steel on the cars that followed.[26]

The Model T's great shakedown trip came in late September 1908. Henry Ford drove one of the prototypes from Detroit to Iron Mountain in Michigan's Upper Peninsula by way of Chicago and Milwaukee, a 1,357-mile round trip. As the Ford Motor Company publication *Ford Times* recounted the story, "The roads going were 6 inches deep in dust—returning after the rains the roads were wet and muddy, and the car when it arrived in Detroit looked as if it had been taking a mud bath." But the most significant problem on the trip was a punctured tire. Gasoline consumption averaged 20 miles per gallon. The Model T's fundamental design was clearly sound. It was time to stop testing and to start building and selling.[27]

# *3* MANUFACTURING THE MODEL T

*A Photographic Tour of Ford's Factories*

At first glance, the meaning of the picture that opens this chapter is not clear. It looks almost like an abstract painting. But it is actually one thousand Model T Ford chassis—the fruit of one nine-hour shift in August 1913—arrayed next to the building in which they were assembled. The picture is full of ironies. It is a publicity shot, taken to illustrate the sheer productive power of the vast Ford factory in Highland Park, Michigan. Yet it doesn't hint at what was really involved in mass production. It is static, but the process it symbolizes was dynamic, characterized by the unceasing flow of materials and parts and a constant search for improved production methods. It is devoid of people, but the plant teemed with workers, some fourteen thousand at the time the photo was taken. The wall is a bland, repetitive backdrop, but it enclosed the most influential industrial plant of the twentieth century.[1]

For all its ironies, the picture is a perfect starting point for a photographic tour about Model T production, because it is a snapshot, a frozen moment, a pause in normally intense activity. Any description of Model T production is necessarily such a snapshot, because the production process was constantly changing. The design of the parts of the car, the materials from which the parts were made, the location of the factories, the layout of buildings within the factories, the layout of machines within the buildings, the design of the machines themselves, the jobs of the workmen—all of these were constantly scrutinized and revised in the Ford Motor Company's relentless drive to lower the cost and raise the rate at which it turned out Model Ts.

Contemporary observers quickly grasped the importance of the production innovations at Highland Park. By 1915 articles on the plant had been published in three major technical periodicals: Oliver J. Abell in *Iron Age,* Fred H. Colvin in *American Machinist,* and Horace L. Arnold and Fay L. Faurote in *Engineering Magazine.* Arnold and Faurote expanded their series into a book, *Ford Methods and the Ford Shops.* Since that time several historians have examined the origins and ramifications of Ford methods in great detail. Most of that work focused on the period of greatest creativity, which ended in 1914. There has been less scrutiny of Model T production methods after 1914. But the Model T stayed in production for another thirteen and a half years, and Ford never ceased making improvements in the production process.[2]

We have already discussed the Model T as an expression of Henry Ford's artistic vision. Others have spoken in artistic terms of Ford's later River Rouge Plant. German engineer Otto Moog said of that plant, "No symphony, no *Eroica,* compares in depth, content, and power to the music that threatened and hammered away at us as we wandered through Ford's workplaces, wanderers overwhelmed by a daring expression of the human spirit." Diego Rivera, who painted famed murals of the Rouge, spoke of "the wonderful symphony which came from his factories where metals were shaped into tools for man's service." But these remarks could apply to the Model T production process as well. It was a work in constant progress, going from (to continue the musical metaphor) a simple song based on a popular melody to a vast symphony that reflected Ford's unique vision. His love of light weight, of simplified work, of ingenious solutions to mechanical problems, and of complete control over the process shaped his factories just as surely as they shaped his automobiles. Allan Nevins and Frank Hill believed that "Highland Park was perhaps the most artistic factory, in architecture, shining cleanliness, and harmonic arrangement, built in America in its day."[3]

Discussing industrial methods and machines can be frustrating, for authors and readers alike. Even knowledgeable engineers can have trouble following written descriptions. Fortunately, historians of Ford are blessed with a rich photographic record, especially for the years after the move to Highland Park. The remainder of this chapter will be driven by that visual record.

The first Ford Motor Company plant. This was a rented frame building on Mack Avenue in Detroit. Owned by Ford stockholder Albert Strelow, it served the company in 1903 and 1904. It was originally a one-story building, but rising sales prompted the addition of a second floor at the end of 1903. Here Ford assembled the Models A and C from purchased parts. Engines and transmissions came from John and Horace Dodge, owners of one of Detroit's premier machine shops. In 1914 the Dodges would go on to build a car of their own. Ford purchased bodies, wheels, tires, and miscellaneous parts from a variety of other suppliers. Each car was completely assembled in one place by one crew, a process known as station assembly. Such methods were no different than those used by other automakers. No photograph of the Mack Avenue Plant assembly operation exists.[4]

Station assembly of Model N Fords, 1906. This photograph was taken on the third floor of Ford's second plant, at the corner of Piquette and Beaubien Avenues in Detroit. Three rows of partially assembled cars are visible. Beyond the cars are rows of Model N engines, stored on their noses to conserve space. The assembly process differed little from the one used at Mack Avenue. The Piquette Avenue Plant, the first built and owned by the company, was three stories tall, 402 feet long, and 52 feet wide. Its brick and wood construction followed the pattern set by New England textile mills. At 63,000 square feet, it was a fraction of the size of the 570,000-square-foot plant Oldsmobile erected in 1903. Ford used the Piquette building from 1904 to 1910. When the company first moved into this structure in 1904, Henry Ford challenged employee Fred Rockleman to a footrace the length of the empty building; history does not record who won the race. Rockleman wondered if the company could ever use all the space. By the time this photo was taken, demand for the Model N made Rockleman's concerns seem quaint.[5]

The home of the Ford Manufacturing Company, 773–775 Bellevue Avenue. We have already seen how Henry Ford used the Ford Manufacturing Company to force Alexander Malcomson out of his way. But the Manufacturing Company also set Ford on the road to mass production. Here Ford engineers learned how to manufacture engines, transmissions, axles, differentials and other parts for the Model N. Material shipments to and from the factory still relied on four-legged horsepower.[6]

Two men were key to the developments at Bellevue Avenue: Walter Flanders and Max Wollering. Flanders sold Ford many of the machine tools used at the Bellevue Plant and recommended that Ford hire the talented twenty-seven-year-old Wollering as superintendent of the new operation. Arriving in the spring of 1906 as the last machines were being installed, Wollering soon had the Bellevue operation humming. In August 1906 Henry Ford hired Flanders to oversee production at both the Motor Company and the Manufacturing Company. Described by Charles Sorensen as a "roistering genius" with "a voice that could be heard in a drop forge plant," Flanders set about rearranging the machine tools at Bellevue Avenue. Typical job shop practice was to group machines by type: all the lathes together, all the milling machines, and so on. This made sense for shops that built small numbers of a large variety of parts. But Ford made large numbers of a small variety of parts. Flanders arranged the tools in order of the operations to be performed. Thus, if a particular part required milling, boring, and turning in that order, a milling machine, a drill press, and a lathe were set up in sequence. Together Flanders and Wollering introduced the Ford people to the advantages of single-purpose machine tools, special jigs and fixtures, and interchangeable parts. Their tenure at Ford was short—both left to join the Wayne Automobile Company in April 1908—but their influence was profound. As Sorensen said, their work "headed us toward mass production."[7]

The expanded Piquette Avenue Plant, 1908. The original three-story brick structure built in 1904 is identified as "The Home of the Celebrated Ford Automobiles." In 1907 the Ford Motor Company absorbed the Ford Manufacturing Company and consolidated all operations at the Piquette Avenue site. The jumbled collection of buildings and parts depicted here reflects the company's struggle to keep up with demand for the Model N. Should the Model T be as successful as Ford hoped, the problems would only grow worse. But Ford managers had a solution in the works.

Highland Park Plant, 1909. In the summer of 1906, the Ford Motor Company purchased a large tract of land in the suburb of Highland Park, just north of Detroit. Albert Kahn, fresh from designing a new factory building for Packard Motor Car Company, was hired as architect. Construction of the new Ford plant began in 1908. In this photo, the first building (designated by the company as Building A) nears completion. It was 865 feet long, 75 feet wide, and featured 50,000 square feet of glass windows.[8]

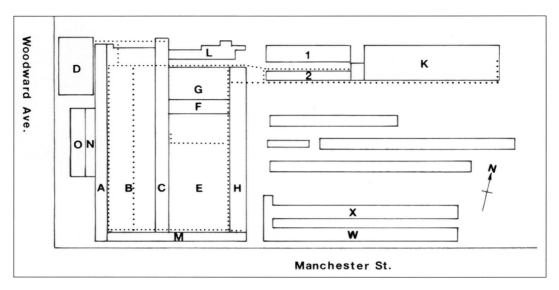

Plan of the Highland Park Plant, 1914. Buildings A, M, and H were four-story factory buildings made of reinforced concrete and faced with brick and large expanses of glass windows. Buildings B and E were one-story machine shops with steel frames and glass sawtooth roofs. Building G was of similar construction and housed annealing furnaces as well as storage for miscellaneous stock material. C and F were glass-roofed crane ways served by electric overhead traveling cranes. The glass windows and roofs flooded the Ford plant with natural light, giving rise to its nickname, the "Crystal Palace." Other key buildings included the foundry K, the power house D, the administration building O, and heat treatment buildings 1 and 2. Buildings W and X were six-story reinforced concrete structures completed in August 1914. The dotted line marks the path of the monorail conveyor system for moving material through the plant.[9]

Overhead monorail conveyor, 1914. This photograph and the next two illustrate how Ford mechanized major material handling. The monorail conveyor was one of the key material handling devices built into Highland Park when the plant opened in 1910. It was a standard system, made by Sprague Electric Works. Electrically powered cars, each driven by an individual workman, pulled trays or platforms carrying various kinds of material. Over one and a half miles of monorail track ran throughout the plant, connecting the foundry, the heat-treating buildings, and all the machine shops. Smooth material flow was critical to the success of the Highland Park operation. Since finished Model Ts weighed 1,200 pounds, turning out 1,000 cars meant moving 1.2 million pounds of material through the plant every shift. The Model T's light weight paid extra dividends here. Had the car weighed 1,700 pounds instead of 1,200, the factory would have had to handle an extra half-million pounds per shift.[10]

Main Highland Park crane way, 1914. This 860-foot-long crane way (marked C on the plan) ran the full length of the machine shops and allowed a pair of five-ton-capacity electric overhead traveling cranes to pick up or deliver material at nearly any point in the process. The photograph shows Model T frames and rear axles stored on the left, with small parts stored in bins on the right. Ford engineers and architect Albert Kahn may have drawn inspiration from a similar crane way at the factory of steam car maker White in Cleveland. A description of White's operation appeared in a magazine article in 1907, while Kahn was designing Highland Park. The photograph is from *American Machinist* magazine.[11]

Mechanization in the Highland Park foundry, 1913. In the center of the photograph is a group of molding machines. Sand was delivered by an overhead conveyor into the bins, which form an inverted V above the machines. Below the bins were molds for cast-iron parts like cylinder heads, pistons, and transmission drums. Workers dropped the sand from the bins into the molds and packed it around patterns for the part to be cast. The patterns were removed, leaving cavities shaped like cylinder heads or other parts. Workers then carried the finished molds to the carousel-like mold conveyor on the left. Each L-shaped pendulum held one mold. The conveyor moved the molds past a ladle that filled them with molten iron. (Molds for engine blocks, larger than molds for the other parts, were laid out on a pouring floor rather than being put on the mold conveyor.) After the iron cooled and solidified, the sand was broken away, leaving the rough castings. The castings moved on to the machine shop via the monorail. Conveyors carried the used sand to a mixing operation and back to the molding machines to make more molds. Ford's molding conveyor was similar to one used by the Westinghouse Airbrake Company in the 1890s. Ford production foreman William Klann said that the sand conveyors were inspired by conveyors that moved grain in breweries.[12]

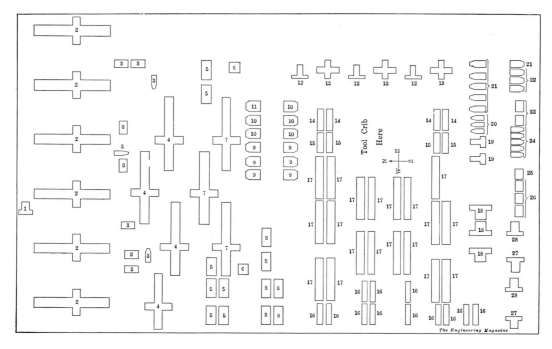

Plan of cylinder block machining operations, 1913. This illustration and the ones to follow demonstrate how fully Ford engineers applied Flanders and Wollering's lessons about the sequential arrangement of machine tools, jigs, fixtures, and interchangeable parts. This diagram is one of the clearest illustrations of the sequential arrangement of machines. Each symbol represents an individual machine tool, and the numbers correspond to the order in which the operations were performed. The monorail transported rough cylinder block castings from the foundry to the north end of the shop in the middle of Building E. The castings then moved south from machine to machine. Twenty-eight separate operations turned the rough castings into finished cylinder blocks.[13]

Cylinder machining Operation 2, milling the bottom of Model T cylinder blocks. Rough engine block castings from the foundry are stacked in the foreground. Workers loaded fifteen blocks at a time into massive Ingersoll milling machines that made two cuts on the bottom of each block. Fixtures held the blocks rigidly in place to ensure accuracy and interchangeability. The machine's horizontal table moved the blocks toward the rear of the photo under the spinning cutters. On the machine at the left, six blocks have passed the cutters, three are still under the cutters, and the six in view have yet to be cut. Two more machines are barely visible to the right, behind the stacks of block castings. A total of six such machines were installed in the shop. Similar Ingersoll millers machined the tops of the cylinder blocks in Operation 4.[14]

Cylinder machining Operation 18, drilling forty-five holes at once from four directions. A fixture held the block rigidly in place and ensured that each block was interchangeable with any other. One of the best examples of a special-purpose machine tool, this device was useless for anything other than making Model T Ford engines. It also required little skill to run. The Ford Motor Company was very proud of this machine and featured it twice in *Ford Times,* the company magazine; this image is from the October 1913 issue. The sort of sequential machining operations used on the engine block were repeated on such parts as engine crankcases, cylinder heads, crankshafts, transmissions, front and rear axles.[15]

Stamping Model T transmission covers, 1912. Several successive stamping operations converted a piece of sheet steel into a transmission cover. Stamping presses were set up in crane way F, and the work moved sequentially from one press to another. Stacks of finished covers are seen at the right. The John R. Keim Company of Buffalo originally supplied Ford with a wide variety of stamped steel parts. In 1911 Ford bought Keim and moved its entire operation to Highland Park.[16]

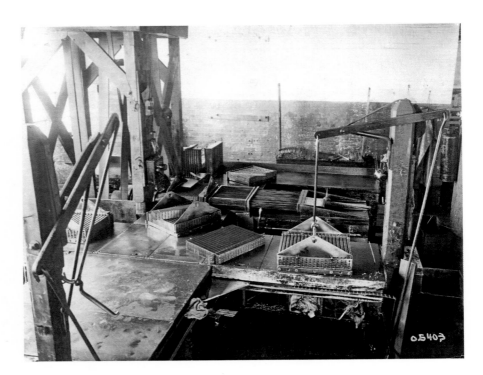

Making radiator cores, 1913. The core of the Model T radiator consisted of ninety-five copper tubes inserted through holes in seventy-four thin brass strips. Workmen placed the tubes and strips in special fixtures and this machine, designed and built by Ford, slid them together in one quick operation. Manned by four workers, the machine assembled a radiator core in two minutes.[17]

Painting spoked wooden wheels, 1911. This image from *Ford Times* illustrated a machine that eliminated the laborious, time-consuming job of hand-painting wheels. The machine dipped a wheel into a vat of paint, withdrew the wheel to a point above the paint but below the edge of the vat, and spun the wheel to remove excess paint. The device was adapted from a drill press, perhaps one that was too worn out to use on precision work. Ford later developed a purpose-built machine for painting wheels.[18]

Dashboard assembly stands, 1913. Ford assembly operations were highly developed before the advent of the moving line. Sub-assemblies like dashboards, axles, engines, and transmissions were put together at benches that were carefully designed for the purpose. Parts were delivered to the benches by hand truck and hand cart and then stored in bins at each bench. Individual workmen then used the parts to make an entire sub-assembly. This photo illustrates the dashboard assembly stands, which were made of wood. The stands are spaced so as to facilitate parts delivery, while the location and arrangement of parts bins are influenced by motion studies.[19]

Engine assembly room, 1913. In normal operation the long tables would have been lined with workmen, each assembling one engine at a time from parts stored in bins on top of the tables and on shelves under the tables. This image appeared in *American Machinist* in 1913.

Station assembly of Model T chassis, 1913. All the various sub-assemblies came together in a chassis assembly process essentially identical to that used at Piquette Avenue. Cars were assembled in groups of fifty. Gangs of workers performed a series of tasks on each chassis, then moved on while another gang performed the next group of tasks. Material was delivered by hand trucks, two of which are seen at the right. The truck nearest the camera holds dashboards, while the smaller truck farther back carries gas tanks. This system was so well orchestrated that a finished chassis left the building every forty seconds. All of these production advances not only came before the moving assembly line—indeed, they made the line possible. Only after the basic infrastructure of material handling and production machinery was in place could the assembly line be developed.[20]

Popular wisdom credits Ford's impressive production increases and price decreases entirely to the advent of the moving assembly line. But this chart tells a different story. The first experiments with putting sub-assemblies like magnetos on a moving line began in April 1913. Experiments on a chassis line began in August 1913. The full effects of the new approach

**Automobile Production, Ford Motor Company, 1908–1917**

| Years | Production | Change (%) | Price (dollars) | Change (%) |
|---|---|---|---|---|
| 1908–1909 | 10,660 |  | 950 |  |
| 1909–1910 | 19,051 | 79 | 780 | −18 |
| 1910–1911 | 34,070 | 79 | 690 | −12 |
| 1911–1912 | 76,150 | 124 | 600 | −13 |
| 1912–1913 | 181,951 | 139 | 550 | −8 |
| 1913–1914 | 264,972 | 46 | 490 | −11 |
| 1914–1915 | 283,161 | 7 | 440 | −10 |
| 1915–1916 | 534,108 | 89 | 360 | −18 |
| 1916–1917 | 785,433 | 47 | 360 | 0 |

Source: *Ford Industries* (Detroit: Ford Motor Company, 1924), 11.

were not felt until the end of the year. But as the chart shows, Model T output increased at a higher rate *before* the advent of the assembly line than after, while the price decreases before and after the line were quite similar. Before 1913, the keys to expanding production and reducing prices were efficient handling of materials, the sequential arrangement of machine tools, the use of fully interchangeable parts, the employment of single-purpose machines, and a highly developed station assembly system.[21]

Chicago meat packing operation, 1915. A conveyor moves hog carcasses past meat cutters, who then remove various pieces of the animal. In their unceasing effort to make Model T production keep up with demand, Ford's engineers borrowed ideas from a variety of places—breweries, air brake factories, steam automobile factories. Sometime in 1913 it dawned on the Ford people that the "disassembly line" principle employed in slaughterhouses could be adapted to their needs.[22]

Flywheel assembly line, 1913. Horace Arnold identified this as the first Ford assembly line. Workers put V-shaped magnets on Model T flywheels to make one half of the flywheel magneto. Simple frames made of pipe and angle iron supported the flywheels. Each worker installed a few parts and pushed the flywheel down the line. The man in the foreground is reaching into the parts bin with his right hand. Productivity gains were immediate. Twenty-nine workers produced flywheels at a rate of one every thirteen minutes and ten seconds per person, compared to twenty minutes per person under the station assembly system. These results were encouraging, but there was plenty of room for improvement. Raising the level of the rails a few inches made the working position more comfortable. Some employees worked quickly, but they had to wait for the slower workers. Moving the flywheels along with a powered chain conveyor forced all workers to adopt the same pace, giving ultimate control of the rate of production to the company. Within a year these and other adjustments allowed fourteen men to produce 1,335 flywheels in an eight-hour shift, a rate of one flywheel every five minutes per person.

Despite the extensive written and pictorial record, there is much confusion about just where the assembly line was first applied at Ford. In a 1953 interview, William Klann, foreman of motor assembly, was adamant that the first assembly line operation was the magneto coil, followed by motor assembly, transmission assembly, and assembly of magnets on the flywheel. But Horace Arnold's account, published in 1915, puts the flywheel magnets first. Given that Klann's recollections of chronology break down when discussing other operations, it is likely that Arnold's identification is the correct one. In the end, the confusion only serves to illustrate how rapidly Ford adopted the new system. As historian David Hounshell wrote, "The development of the assembly line at Ford was so swift and powerful that it defied accurate, unambiguous, timely documentation by the Ford Motor Company and its employees."[23]

Motor assembly line, 1914. After the success of the magneto assembly line, experiments began with transmission and motor assembly. Here workers install pistons and connecting rods on the motor line. In addition to pure assembly operations, the motor line also included some manufacturing tasks such as pouring babbitt bearings and finishing valve seats. Ford engineers continuously refined the process, eventually lowering engine assembly time from 594 man-minutes to 226 man-minutes.[24]

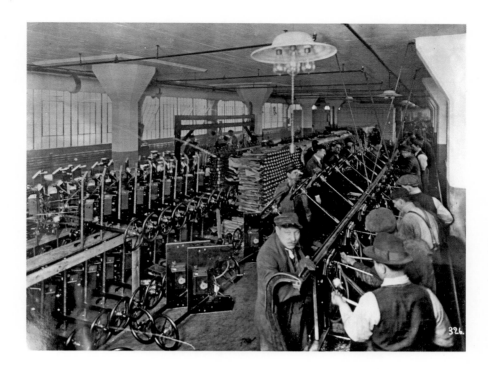

Dashboard assembly line, 1914. The assembly line tidal wave rapidly rolled through the Highland Park Plant, sweeping all before it. Compare this view of the dashboard line with the picture of the dashboard assembly stands on page 48. Within months the carefully planned station assembly stands were consigned to the scrap heap.[25]

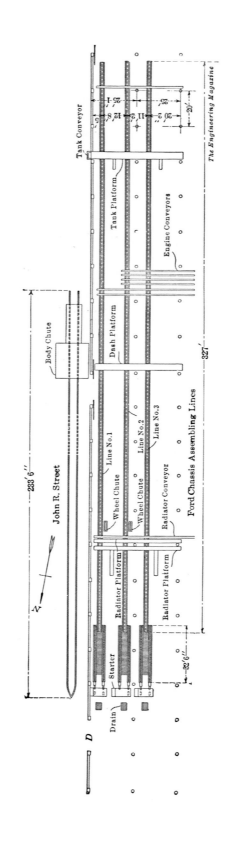

*(left)* Plan of the chassis assembly lines, 1914. The process moved from south to north in Building H. On the chassis lines, frames, axles, gas tanks, engines, dashboards, wheels, radiators, and bodies came together in that order to produce finished, running automobiles. Ford's experiments with a chassis line began in August 1913 and continued into the following year. Engineers tried different numbers of workmen and different timings of material delivery. Sometimes they pulled the chassis along by a rope and a windlass. Sometimes they pushed the chassis along on its own wheels. By early 1914 a chain-driven line was in operation. By mid-1914 three parallel lines were in operation, as shown here. The next five photographs illustrate the major assembly operations.[26]

Gas tank installation, 1914. Chassis assembly began with the attachment of the front and rear springs and axles to the frame. Then came the installation of the gas tank, as shown here. Prior to installation, tanks were filled with a gallon of gas so that the car could be started at the end of the line. At the time this photo was taken, two versions of the moving line were in use. In the foreground workers still pushed cars on their wheels by hand, while in the background a moving chain is in use.[27]

Engine installation, 1913. Conveyors moved the assembled engines eastward from Building E. Workers then used chain hoists to lower the engines into place on the chassis. The worker at the front of the car is guiding the engine down, while the worker at the middle of the car is already connecting the driveshaft to the transmission. Neither of the two assembly lines visible in this heavily retouched photo has yet to be put on the moving chain.

Installing dashboards, 1915. Assembled dashboards were conveyed to the overhead platform, as shown in the back of the photo, and slid down a ramp to the assembly line. Workers then connected ignition wires, spark controls, and throttle controls to the engine, and connected the steering column to the tie rods on the front axle. By this time, all three assembly lines are on the moving chain.[28]

Radiator and wheel installation, 1914. As with gas tanks and dashboards, radiators were conveyed to an overhead platform and slid down ramps to the line. Wheels, with tires mounted and inflated, rolled down chutes from the upper levels of Building H. By the time this photograph was taken, a Model T could be assembled in a bit more than one and a half man-hours, compared to twelve and a half man-hours needed when using station assembly.[29]

Starting the engine, 1914. At the end of the line, the rear wheels of the finished chassis dropped over a set of powered rollers in the floor. The rollers spun the wheels to start the car. In front of the car, a worker fills the radiator with water. This car would be packed in a railroad car and shipped separately from its body, as shown in the next photograph. Just behind the car being started is a touring car with its body installed. The front edge of its cloth top is just visible at the right hand side of the photo.[30]

Temporary body installation, 1913. This misunderstood photo does *not* depict the body drop where body and chassis finally came together. A 1915 publication called *Factory Facts from Ford* explained the scene: "The bodies are at this time placed on the chassis merely as a means of a rapid transportation to the freight cars, for in ordinary transportation the bodies are packed in the cars separate from the chassis." At the loading dock the body and wheels were removed and packed separately to save space. Cars to be delivered fully assembled had their bodies installed inside the plant near the end of the assembly line.[31]

Crowd gathered on Manchester Street outside Building M after the announcement of the five-dollar day, 1914. As Ford production men surveyed the phenomenal productivity increases of their new assembly line methods, they also noted a disturbing by-product: the workmen disliked the system and expressed their displeasure by taking jobs elsewhere. By late 1913, even before assembly lines were fully implemented throughout the shop, labor turnover was a whopping 380 percent. People quit so often that in order to expand the labor force by 100 men, the company had to hire 963. In January 1914, Ford—probably in response to the urging of James Couzens—took perhaps his boldest step ever, one that transcended engineering. The company announced that it would pay five dollars for an eight-hour day. Since the previous rate had been $2.34 for a nine-hour day, this was a shocking announcement. The work in the factory was no easier, the pace no less relentless. But the pay was now so good that people were willing to do the work. In practice the new wage went only to people deemed "qualified" after an investigation into their private lives. Nevertheless, the lure of the money was so strong that most employees put up with such paternalistic policies, however reluctantly. This bargain between Ford and his workers—submission to the relentless discipline of the line in return for high wages—would turn out to be as important as the Model T itself.[32]

New Shop at Highland Park, 1914. To outside observers, Highland Park was an industrial wonder. But to Ford engineers the plant was increasingly becoming inadequate. Despite the monorail and crane ways, the company still employed more than a thousand men to truck, push, and drag material through the various buildings. Despite the engineers' best efforts, the arrangement of machines and departments was not optimal. And despite the plant's impressive size, it was not big enough to accommodate the increasing demand for Model Ts. In 1913 Ford began designing four massive new six-story additions, to be 842 feet long and 62 feet wide. Between each pair of buildings, the engineers placed a glass-roofed crane way. Erected at the southeast end of the Highland Park complex, the first two buildings (lettered W and X in the plan view on page 38) opened in August 1914. In this image from *Ford Methods and the Ford Shops,* the roof gables mark the locations of the crane ways. By 1917 two similar buildings (lettered Y and Z) were in place parallel to these.[33]

Looking east, down the crane way between Buildings W and X, 1914. Material arriving by rail was easily delivered to any level of either building. The new buildings allowed Ford to move painting and upholstering of automobile bodies in-house. For many years Ford bought its bodies from outside suppliers. As late as April 1914, 95 percent of those bodies came to Highland Park already painted and upholstered. By October, 40 percent of Model T touring car bodies were being painted and trimmed in the new buildings. In this photograph, a number of painted touring car bodies can be seen drying on the right side of the crane way, near the platform labeled W5-9.[34]

Touring car bodies arriving, 1915. Purchased touring car bodies, unpainted and un-upholstered, came by horse-drawn truck from Briggs Manufacturing to the west end of Building W. Near the center of the photo is an inclined conveyor, carrying the bodies to the fifth floor.[35]

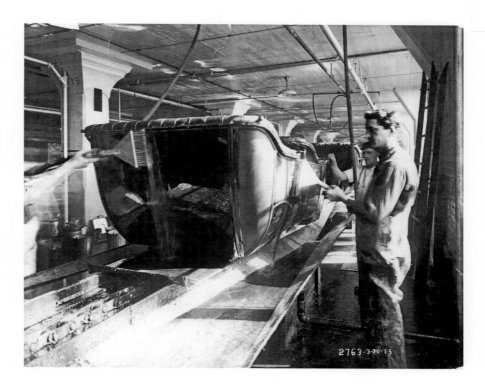

Final painting of bodies, 1914. Unpainted bodies received a primer coat and two color coats that required twenty-four hours drying time between each coat. The bodies then traveled to a moving line where seats and upholstery were installed. Upholstered bodies received one final coat of clear varnish, as shown here. Workers "flowed" the varnish on with applicators that looked like vacuum cleaner tools. Excess varnish dripped off into catch trays to be recycled.

Contrary to popular opinion, not all Model Ts were black, only 11.5 million of them (out of 15 million total). Early 1909 cars were red, gray, or dark green; in mid-1909 all became dark green; early in 1911, all became dark blue; the blue prevailed until late 1914 or early 1915, when black became standard. (It should be noted that the greens and blues were so dark that they often appeared black, especially in photographs.) In August 1925, with Model T sales falling, Ford re-introduced colors.

Black was *not* chosen because it dried faster than other colors. There is no evidence that black dried any faster than any other dark varnishes used at the time for painting cars. Rather, black seems to have been chosen by Ford because it was more durable and less expensive than colored varnishes. Even after the reintroduction of body colors, high-wear areas like fenders and running boards continued to be painted with durable black.[36]

Installing folding tops, 1915. After the final varnish coat dried, folding tops were installed on yet another moving line.[37]

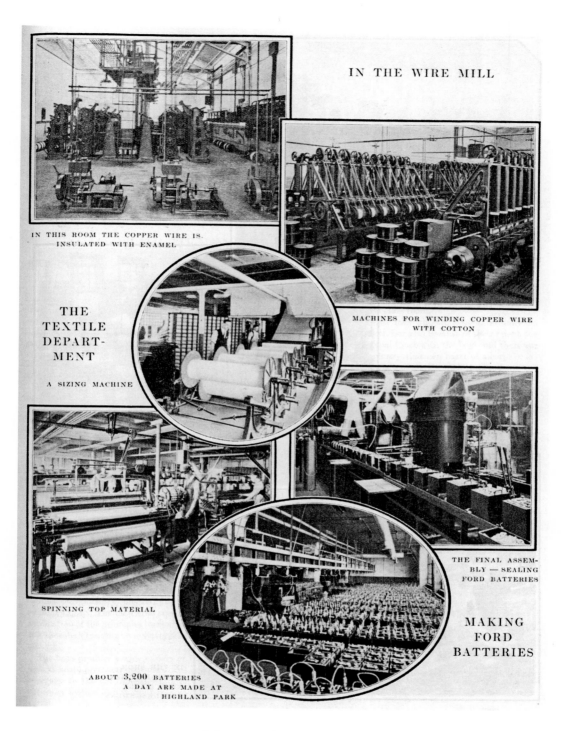

IN THE WIRE MILL

IN THIS ROOM THE COPPER WIRE IS INSULATED WITH ENAMEL

MACHINES FOR WINDING COPPER WIRE WITH COTTON

THE TEXTILE DEPARTMENT

A SIZING MACHINE

SPINNING TOP MATERIAL

THE FINAL ASSEMBLY — SEALING FORD BATTERIES

MAKING FORD BATTERIES

ABOUT 3,200 BATTERIES A DAY ARE MADE AT HIGHLAND PARK

Other manufacturing operations, 1924. As a means of controlling both cost and quality, Ford began to manufacture more materials he had once purchased. Facilities were added for making windshield glass, cloth for tops, artificial leather that waterproofed the tops, plastic (called Fordite) for steering wheel rims, rear axle bearings, wire for ignition systems, even storage batteries (after electric starters became available on Model Ts). This 1924 Ford brochure illustrates many of the diverse operations added to Highland Park over the years.[38]

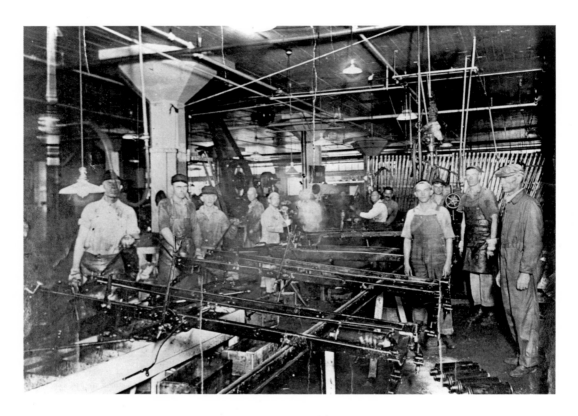

Branch plant, 1917. The Highland Park Plant was so well documented and so famous that it obscured the vital role played by Ford's branch plants. In 1910, Ford built its first branch assembly plant in Kansas City, Missouri, where this photograph was taken. This move allowed the company to reduce costs by shipping parts rather than whole cars. By 1915 there were assembly plants in twenty-eight cities. Highland Park produced 1,200 complete cars each day, while the branches assembled a total of 2,500 cars per day. In 1921 the company began a huge project to modernize and expand the branches. By 1925 there were thirty-six branches and the revitalization program had cost $125 million.[39]

Branch plant, Manchester, England, 1926. The attributes that made the Model T such a success in the United States—low cost, light weight, simplicity, reliability, the ability to traverse bad roads—were just as attractive overseas and would make the "T" the first "world car." But Ford Motor's foreign expansion predated the Model T. In August 1904, the Ford Motor Company of Canada was established as a separate company with a majority of its stock controlled by the American company's stockholders. At a plant in Windsor, on the Canadian side of the Detroit River, cars were assembled from a mix of American and Canadian parts, and the new company met with modest but not overwhelming success. In the meantime, the Ford Motor Company explored the export market; by 1907 it was selling small numbers of cars in Great Britain (103 cars), Germany (41), Belgium (24), and Mexico (23).[40]

The Model T, of course, changed everything. With a superior product to sell, Ford expanded rapidly. The Windsor plant gradually moved from assembly to manufacture, and the Canadian company handled distribution to all parts of the British Empire except Great Britain itself. Assembly plants went up all over the world: England in 1911; France in 1913; Argentina in 1916; Denmark in 1919; Spain and Uruguay in 1920; Italy and Belgium in 1922; South Africa in 1924; Japan, Australia, and Mexico in 1925; and Germany, Malaya, and India in 1926. The distribution process mirrored the one developed in the

United States. Cars were shipped "knocked down"—that is, partially assembled—from Detroit or Canada. Final assembly took place at the foreign branches, which also served as distribution points for completed cars. Like the Windsor plant, the Manchester operation gradually added more and more manufacturing processes.[41]

Rouge Plant, 1924. As impressive as the Highland Park Plant was, its potential was limited. The 56-acre site could not be easily enlarged, and most of the buildings were not designed with mass production in mind. The experiences of Ford engineers during their manufacturing revolution led them to overhaul their ideas about plant architecture and layout. For his next great composition, Henry Ford would draw on all his past experience to go beyond the Highland Park symphony and create a true magnum opus, an industrial grand opera.

On a thousand acres of marshy ground three miles from where the Rouge River merged with the Detroit River, Ford began building a vast plant intended to take the idea of flow to its logical conclusion. Here raw materials—iron ore, coal, limestone—would flow through the plant, being transformed into iron and steel, which in turn would become engines, transmissions, bodies, and eventually complete cars.

Construction began in 1917, and as work proceeded many Model T operations were gradually transferred to the new site. By 1926 the Rouge had blast furnaces, coke ovens, steel furnaces, and rolling mills. A stamping plant produced Model T body panels, while another plant built complete bodies and parts that were shipped to the branches for assembly. The world's largest foundry made engine castings and fed a new building dedicated to engine manufacturing and assembly next door. A glass plant vastly larger than Highland Park's made window and windshield glass. The various processes were tied together by a network of monorails, tramways, conveyors, and railroads that dwarfed the systems that had amazed observers at Highland Park.[42]

The end of the Model T, 1927. No doubt Ford planned to move final assembly of Model Ts to the Rouge, but events intervened. Model T sales and production peaked in 1923, declining steadily thereafter. The Model T's virtues of ruggedness and simplicity no longer satisfied buyers, who wanted comfort, style, and speed as well. General Motors and upstart Chrysler were more in tune with these changing tastes; Ford needed a new design to compete. On May 26, 1927, Edsel Ford drove a pensive Henry out of the Highland Park Plant in the fifteen-millionth Model T. An additional 458,781 cars would be assembled in various branch plants, but this was the symbolic end of Model T production.[43]

# 4 SELLING THE MODEL T

"You do not sell goods, but ideas about goods." So wrote Norval Hawkins, who for a dozen energetic, creative years was sales manager of the Ford Motor Company. Hawkins sold a very complex set of ideas about the Model T. The public bought those ideas and the car that embodied them at ever-increasing rates for a decade and a half. But while public expectations changed, the Model T remained fundamentally the same.[1] Over its nineteen-year run, the Model T underwent thousands of detail changes. Styling was updated several times, electric lights and eventually electric starting were added, and nearly every individual part was altered in the continuing effort to lower costs and increase production. But the design of significant elements—the suspension system, the transmission, the magneto, the torque tube drive, the 4-cylinder engine—never fundamentally changed. Gradually the car ceased to embody the ideas that once defined it, and it lost its appeal. By then Hawkins was long gone, out of the line of fire as Henry Ford sought to understand how and why the world had changed.[2]

When Hawkins arrived at Ford in 1907, he found a solid foundation on which to build an expanded sales organization. Using the auto shows held in major cities as a recruiting ground, James Couzens built up a good network of dealers. By 1905 some four hundred fifty Ford agencies, usually headed by men personally selected by Couzens, were in place. Ford taglines and slogans like "Boss of the Road," "The Car of Distinction," and "Watch the Fords Go By" were all implemented on his watch. The company also boasted a memorable logo, the Ford script. It was contributed by Harold Wills, who retained from his boyhood a printing set that included a script typeface. He set the company's name in that typeface, and an enduring corporate symbol was born.[3]

In addition, Hawkins inherited two powerful ideas to sell.

The first was the company's image as an underdog, fighting the grasping "trust" that held the Selden patent.

For the Ford Motor Company, the Selden patent suit was a classic case of being handed lemons and making lemonade. The Selden patent was the handiwork of Rochester, New York, lawyer George Selden, who designed (but did not actually build) a horseless carriage with a gasoline-burning internal combustion engine. He filed for a patent on his idea in 1879, but for the next sixteen years he delayed final approval by regularly filing minor modifications to the design. In 1895, sensing that a viable horseless carriage industry was about to be born, Selden finally permitted the Patent Office to act on his application. In November he received a patent that he claimed covered all gasoline-powered vehicles designed since his original 1879 application and all that would be designed, built, and sold in the United States until the patent's expiration in 1912. Having no interest in actually making a car, he assigned the patent to the Columbia and Electric Vehicle Company in 1899, which reorganized as the Electric Vehicle Company in 1900. The Electric Vehicle Company filed patent infringement suits against five automobile makers. Ultimately these five, along with ten others, settled with the company and agreed to form a patent-pooling combination called the Association of Licensed Automobile Manufacturers. ALAM collected royalties on each car sold (paying a portion to Selden and the Electric Vehicle Company) and decided which automakers could gain admission to ALAM. ALAM would sue any manufacturer not admitted and run it out of business.[4]

The Ford Motor Company applied for ALAM membership in 1903, but the association rejected the application on the grounds that Ford was merely an assembler, not a manufacturer. ALAM then published an ad in the *Detroit News* warning that makers, sellers, and buyers of unlicensed cars could be prosecuted by ALAM. Ford published a counter advertisement in the *Detroit Free Press* promising protection for its dealers and customers, and the war was on. The antagonists traded barbs in paid ads until ALAM sued in late 1903. The case dragged on for over seven years, with Ford losing in 1909 but ultimately winning on appeal on January 9, 1911. Ford's fight with ALAM coincided with the rising tide of Progressivism and public concern with the growing power of big business. The Selden patent suit came to be seen by the public as a battle of the little guy against the big bully. It marked the first time that the national media and the public at large noticed the Ford Motor Company and its founder.[5]

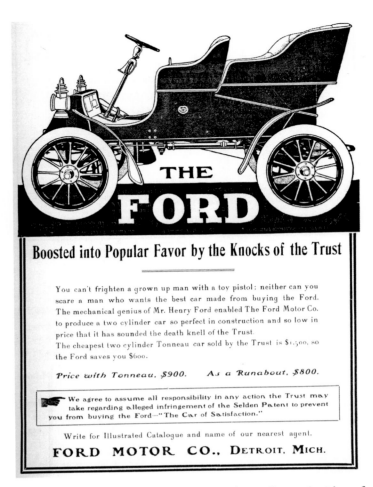

A 1904 Ford ad casts Ford as the defender of the little guy against the grasping ALAM "Trust."

The second great idea Hawkins had to sell was the idea of Henry Ford. Ford was hardly unusual in naming his company and his product after himself, but he proved particularly adept at personalizing both his car and his business. When Ford set a new automotive speed record in 1904, ads trumpeted his success, noting that "it is not uninteresting that the builder and driver of this car is also the designer and builder of the regular Ford Runabout and Tonneau made by The Ford Motor Co., Detroit, Mich." Another ad from 1904 touted the Ford car as being "Modeled by the Master Hand of America's Foremost Automobile Designer." In 1906, a series of pithy quotes appeared in issues of *Horseless Age*. Lines like "They criticize— but they copy while they criticize" and "The man who has always been right in the past can safely be trusted for the future" were followed by the single word *Ford*. The clear implication was that these words had come straight from Henry's lips. It also didn't hurt that the Ford script logo looked enough like

> # One Mile in 39⅖ Seconds
>
> Was made at Detroit last Tuesday on a straightaway course, this speed being
>
> # 91⅓ MILES AN HOUR
>
> This wonderful ride was made by Mr. Henry Ford, on that old reliable, unequaled speed machine, The
>
> # FORD
>
> 999 Racer. Being officially timed under the rules of the American Automobile Association, this new figure is the
>
> # WORLD'S RECORD
>
> for a straightaway mile, any class of car, breaking the previous mark, held in France, by 6 3-5 seconds.
>
> It is not uninteresting that the builder and driver of this car is also the designer and builder of the regular Ford Runabout and Tonneau made by
>
> **The Ford Motor Co., Detroit, Mich.**

*Ford's 1904 land speed record, set on the frozen surface of Lake St. Clair north of Detroit, raised the profile of both the man and the car.*

Henry's own signature to make it seem that he was "signing" every car himself.[6]

Norval Hawkins is a fascinating example of a man who made a great recovery from a serious ethical tumble. In 1894 he was convicted of embezzling $8,000 from his employer and served a prison sentence. But influential friends held him in such regard that they came to his aid after his release. Amazingly, he set up a successful accounting firm and came to Henry Ford's attention while auditing the Ford Motor Company's books. Ford employee George Brown, who described Hawkins as having "a wonderful set of brains," said that when Hawkins arrived at Ford Motor Company he "revolutionized the old sales division.… He just turned things topsy-turvy and everything seemed to thrive. He had something new in salesmanship."[7]

Ford sold its cars through a dual system of company-owned branches and independent agencies. Branches were located in large cities, while agencies served smaller cities and towns. Branches also served as distribution points, to which cars were shipped and from which agencies picked cars up. Hawkins expanded the branch system, locating new ones in cities where railroad freight rates changed in order to minimize shipping costs. A cadre of "road men" traveled to both branches and agencies, inspecting their books and their physical facilities and reporting back to the home office. Dealers were assigned well-defined territories, and underperforming dealerships always faced the threat of having their territory narrowed.[8]

Hawkins had no interest in dealers who simply sat back and waited for customers to come in; he demanded active selling. He expected dealers to canvass their territories to identify prospects, to keep careful records of their contacts, and to call regularly on their prospects. Hawkins also believed in hands-on management. He often visited agencies himself and demonstrated selling techniques by calling on potential customers.[9]

One of Hawkins's brainstorms was the company magazine, *Ford Times*. First issued on April 15, 1908, it was filled with information on car design, production methods, testimonials from owners, stories of Ford victories in races and hill climbs, and advice and encouragement for dealers. Lively and well illustrated, *Ford Times* boosted morale at the agencies and branches. It was also sent to any existing or potential Ford owner who requested it, thus serving as another advertising medium.[10]

On March 18, 1908, a month before the appearance of *Ford Times,* another important publication was sent out to Ford dealers and branches. It was the *Advance Catalog* for the Model T. Following Henry Ford's vision, the car's creators targeted a market that they *believed* was out there. The response to the catalog confirmed their belief. Typical was a comment from a New Castle, Pennsylvania, dealer who wrote, "It is without doubt the greatest creation in automobiles ever placed before a people, and it means that this circular alone will flood your factory with orders." The orders did indeed flood in, by telephone, telegraph, and mail, even though deliveries were not scheduled until October 1. In September Couzens brought the managers of the fourteen company-owned branches to Detroit for a briefing on business conditions and on the company's plans to produce twenty-five thousand of the new cars in the coming year. The managers responded by demanding fifteen thousand cars for the branches alone.[11]

Once the cars appeared in dealerships, the demand rolled on in a continuous wave. By May 1909 Ford announced that it would temporarily stop accepting new orders because every car

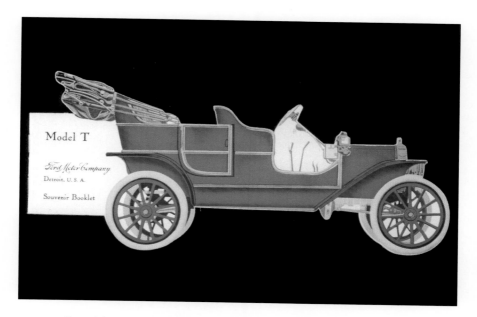

One of the earliest Model T sales brochures is this clever booklet in the shape of a 1909 touring car. Note the "two-lever, two-pedal" control system used on only the first 500 cars. Red was a standard Ford color in 1909.

Ford's industrial wonder at Highland Park was a popular image for postcards like this one from 1914. During the summer the factory's huge expanses of glass windows were shaded by huge expanses of canvas awnings.

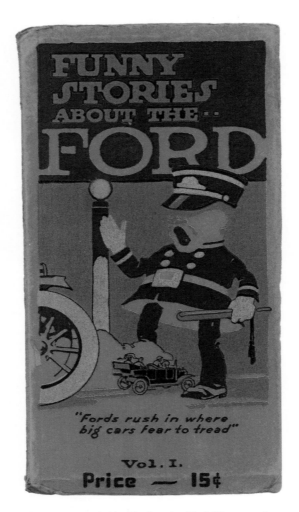

The homely, ubiquitous, reliable, life-changing Model T spawned a new humor genre—the Ford joke. Enterprising publishers cashed in on the phenomenon by offering collections of Ford stories in inexpensive little booklets like this one from 1915.

This piece of 1924 Ford sales literature featured a somewhat surreal illustration of a Ford-owning family bursting through the page to meet up with another family of Ford drivers.

An ad from 1924 trumpets a theme that still resonates with consumers today—the automobile as a symbol of freedom.

Despite this 1924 advertisement's declaration that Model T demand was "wide and ever-growing," production had in fact peaked in 1923 and fell every year thereafter, until the car was discontinued in 1927.

Henry Ford's resistance to selling Model Ts on credit created opportunities for other people. This 1925 poster advertised a finance company that made loans to Ford buyers.

By the time this ad appeared in 1925, the Model T's technology and styling were both outdated, so Ford chose to emphasize the car's reliability, which could not be denied.

This scene of a farm family visiting neighbors illustrates one of the Model T's most important effects—breaking down the isolation of rural life. It is one of eight paintings Norman Rockwell created in 1952 for Ford Motor Company's fiftieth anniversary in 1953.

# FORD

always be relied upon to present honestly the merits and demerits of any proposition. Where the Ford buyer is absolutely safe is in the fact that he is buying a car designed by the most successful, capable and best known automobile engineer in the world,—Henry Ford. In the further fact that he is dealing with a company which has actually built more cars than any other company and that there are today 16,000 Ford cars proving Ford superiority. In buying a Ford, you buy a car with a reputation for quality second to none, from a company that considers reputation its biggest asset and the maintenance of it of chief importance.

The Model "T" touring car offers the greatest automobile value ever announced by the Ford Motor Company and that means the greatest value ever offered for the Ford has always led the procession. A careful perusal of the following pages will convince even the most skeptical that it is years ahead of any other car in design.

The price is remarkably low—so low that you will wonder how it is accomplished and may even doubt the quality. When Ford announced the price on his now famous runabout, skeptics said "impossible," but the car made good and an enormous number were sold, the very thing we had figured on and the thing that made the price possible. The same quantity production methods that made it profitable to sell the runabout at a figure lower by one-half than a car of similar specifications had ever been sold for, will be utilized to keep down the cost of Model "T." At that there is less profit per car by a considerable margin than is usually figured in the selling price of most cars, but half the ordinary profit, multiplied by four times the number of sales still gives us 100 per cent. more profit than the other fellow.

### Model "T" Features

Steering gear and control on left side of car.

Engine, transmission, fly-wheel, magneto and universal joint enclosed in same case.

Top of engine removable so that the valves, cylinders, etc., can be readily cleaned, repaired or adjusted.

With high speed in, any speed quickly obtainable, from a dead stop to 40 miles an hour by operating foot lever.

Magneto is a part of fly-wheel, is a miniature alternating current generator of Ford's own design.

Splash system of lubrication with the flywheel as the distributing agent.

Ford Vanadium steel throughout—this steel

# FORD

Right Side—Model T

made by us from our own analyses and guaranteed to possess greater dynamic qualities than any other known steel.

Simplicity of design and construction, fewer parts, more easily adjusted and repaired than any other car manufactured.

Not an ounce of dead weight—plenty of weight to meet every conceivable demand.

Price, $850.00 F. O. B., Detroit, U. S. A.

### Motor

The Model "T" Touring Car is equipped with a four-cylinder vertical engine rated at 20 H. P. Size of cylinders, 3¾x4. Cylinders of finest quality gray iron.

Some of the noteworthy features found in the Model "T" engine are:

By removing twelve bolts the entire top can be taken off exposing all four cylinders, all four pistons and all eight valves. If it is desired to clean cylinders, valves, etc., a thorough job can be quickly done, valve surfaces ground if necessary, and top replaced.

The crank case is oil tight and in addition to enclosing the crank shaft, forms the lower half of the housing of the transmission, fly-wheel, magneto and flexible joint, all of which are enclosed and operated in an oil bath. This form of construction makes dripping of oil impossible as all working parts are enclosed. The fly-wheel is back of the engine and is also utilized as a rotor for the magneto.

Crank and cam shafts are drop forged each of a single piece of Ford Vanadium steel heat treated after forging, all bearing surfaces ground to absolute accuracy.

Connecting rods are drop forged from Ford Vanadium steel.

The Commutator is in front, easily accessible

### Three Point Suspension

In the Model "T" touring car the Ford plan of 3 point suspension so successful in other Ford models has been utilized. The motor is 3 point; the front axle is 3 point; the rear axle is

A page from the 1908 *Advance Catalog* sent to dealers. It emphasized light weight, simplicity, the three-point suspension, and low price.

scheduled for production through July was already sold. It did not resume taking orders until July 7. The first Model Ts sold themselves, but Ford Motor Company mounted a full-scale effort to keep demand growing.[12]

Many of the early Model T advertisements and brochures were quite typical of the era, filled with text describing the technical features of the car, extolling the virtues of vanadium steel, and emphasizing the vehicle's low price. They touted the Model T's advantages *over* other cars, and thus were aimed primarily at people who had already made up their minds to buy a car and were trying to decide which one. For many people the Model T was the only real option because it was the only car they could afford. But the real key to continuing Ford's sales into the future was reaching people who *hadn't* decided to buy a car. They had to be persuaded that car ownership would pro-

A typical early (1908) Model T ad emphasizing technology and price.

vide them with something they wanted but didn't already have. It was with these potential customers that Hawkins's insights about the non-rational side of salesmanship came into play. The sales appeal, said Hawkins, "must be made primarily to the heart instead of the mind. A man's emotions, not his thoughts, control his Desires."[13] A 1910 article in *Ford Times* brought this approach home to dealers. Under the title "Why Doesn't More Auto Copy Talk My Language?" the author wrote, "I am quite sure now I wish to have nothing to do with a car's mechanism—I am a joy rider pure and simple. The time is now come for automobiles to be advertised as a necessity to one's health and comfort, and the pleasure which they give. The automobile is a necessity—the world was ready for it and embraced it when it came or else it could never have made such wonderful headway."[14] A survey of Ford sales materials reveals that they targeted many different emotions, often in the same ad. However, certain ideas appear again and again in different guises.

Ford told potential buyers that their purchase put them in the vanguard of modern life. A 1911 brochure linked Henry Ford with the era's greatest symbol of technological progress,

Thomas Edison. The inventor of recorded sound, moving pictures, and electric lights identifies Ford as "one of a group of men who has helped to make the United States of America the most progressive nation in the world." A pamphlet from 1910 urges people to "Buy a Ford Car, because when you do, you are in the forefront of automobile advancement. A Ford Car is years ahead of every other car offered at this time." Even the typical technological appeals were couched in the language of modernity: "To-day's light, strong, Vanadium-built Ford is tomorrow's car." Doctors, for whom house calls were still a routine part of life, were urged to add the new transportation technology to their arsenal of disease-fighting weapons: "When minutes mean life and death—as they do in almost every day's work for the doctor—the possession of a Ford car becomes to the physician an imperative demand for humanity's sake."[15]

Advertisements repeatedly emphasized the twin themes of comfort and pleasure. One spoke of the simple comfort of riding in a warm, dry Ford coupe as opposed to getting wet and cold waiting for a trolley car and then standing "in the crowded car on the wet floor while the cold breezes chase the dangerous chills up and down your back every time the door opens." Another spoke of the democratization of pleasure made possible by Ford's vision "that the automobile, like the piano, might be made a commodity, something every family might own and enjoy." A 1913 ad expressed similar sentiments: "If there were no Fords, automobiling would be like yachting—the sport of rich men. But by centering his effort upon the production of one good car, Henry Ford has brought the price down within reason—and the easy reach of the many."[16]

The Ford Motor Company was not above discreet appeals to romantic pleasure. A 1911 pamphlet shows a well-dressed man assisting an equally well-dressed lady as she alights from a Ford coupe. The caption reads, "A very pretty girl and a charming scene from California." A photo in *Ford Times* depicts a Model T touring car, its back seat occupied by two young ladies and one lucky young man. The caption goes through three rounds of "He loves me—he loves me not," before concluding, "Oh pshaw, we haven't got time to finish this. Look at the picture and decide for yourselves." The lucky male who owns a Ford, the ad implies, will have his pick of pretty girls.[17]

Sometimes the appeal to pleasure was far more direct. "Obey that impulse!" commands a Ford ad in a 1914 issue of *Horseless Age*. "The fine joy of automobile ownership may now be yours. Ford prices are down within the easy reach of the untold thousands who have waited for the coming of the right car at the right price."[18]

*A very pretty girl and a charming scene from California*

As this 1911 ad demonstrates, there is nothing new about promising that the right car will attract pretty girls.

Hawkins singled out the women's market for special attention and made surprisingly explicit appeals to women's growing sense of liberation from nineteenth-century strictures. A 1911 booklet entitled *The Lady and Her Motor Car* opens with a purported testimonial from an enthusiastic female customer. In her Ford, she wrote, "I take my fill of the summer sunshine, the blue skies, and I know the brilliant woods of autumn as I never did before. I have found health, vigor, contentment, diversion, confidence in myself and a source of never ending, ever varying pleasure in driving my car." Four years later an updated booklet called *The Woman and the Ford* is even clearer. "It's woman's day," the text declares. "No longer a 'shut in,' she reaches for an ever wider sphere of action—that she may be more the woman." "The car," the brochure goes on, "is a real weapon in the changing order."[19]

Conventional advertising was not the only arrow in the Ford Motor Company's quiver. In 1909 an opportunity presented itself in the form of a transcontinental auto race from New York to Seattle. Several makers declared their intention to

82 / THE MODEL T

participate, but only six cars actually competed: two Fords (each with 20 horsepower, each weighing 1,200 pounds), a Stearns (46 hp, 4,600 lbs), an Acme (48 hp, 3,500 lbs.), a Shawmut (45 hp, 4,500 lbs.), and an Itala (50–60 hp, 4,000 lbs.). Henry Ford could not have designed a better test of his belief in the virtues of lightweight automobiles. The rugged little Fords proved to be perfectly suited to the wretched American roads. The Model T wearing number 2 covered the 4,100-mile route in 20 days and 52 minutes, finishing 17 hours ahead of the second-place Shawmut. Ford trumpeted the victory in many national publications and published a booklet called *The Story of the Race*. The winning car even made a return drive to New York, stopping at dealerships along the way for promotional events. Unfortunately, five months later it was revealed that Ford mechanics had illegally replaced the Number 2 car's engine during the race, making the Shawmut the real winner. Henry Ford didn't care; he had already banked the publicity dividends.[20]

Track racing also helped sell Model Ts. Ford's factory driver Frank Kulick and a phalanx of privateers made stock and modified Model Ts feared competitors at the smaller dirt ovals across the country, regularly winning against larger, more expensive cars. Racing rules often discriminated against the light Model T, requiring drivers to add ballast to meet minimum weight requirements. Protesting such rules, Ford in 1911 withdrew its factory support for racing, but private owners continued to race Model Ts successfully.[21]

In 1975, the Chrysler Corporation sought to revive sagging sales by rebating part of the purchase price of a new car back to the customer. Company spokesman Joe Garagiola introduced the policy to television audiences with the line, "Buy a car, get a check!" Chrysler may have thought it was on to something completely new, but it was over sixty years behind Henry Ford. Henry, however, wasn't trying to revive sales. He simply wanted to keep sales going and to reinforce his company's image as the friend of the little guy. On July 31, 1914, the Ford Motor Company offered to give rebates of between forty and sixty dollars to each Model T buyer if sales exceeded three hundred thousand cars during the following year. On August 1, 1915, came the announcement that Ford sales had totaled 308,213 during the previous year and that each buyer would indeed "get a check" for fifty dollars. Ford Motor Company's bill came to $15,410,650. One observer called these dollars "the most virile crop of good will seeds ever planted."[22]

For all its innovative sales techniques, the Ford Motor Company deliberately rejected the most important marketing development in the history of the auto industry—credit. In the early years of the industry, car manufacturers sold to dealers for cash and dealers sold to customers for cash. Dealers might be able to arrange bank loans to finance their purchases from the car company, but banks did not consider it sound practice to loan money to individuals for purchasing a car. This made buying even an inexpensive car like the Model T a challenge. For instance, in 1916, when the price of a Ford had dropped to $360, that sum still represented nearly half a year's income for the average blue-collar worker.[23]

Private owners looking to dispose of used cars seem to have pioneered the practice of selling cars on credit. By 1909 a typical classified ad in the *Chicago Tribune* offered a used Overland for "$800 cash, balance monthly," while another notice in the same paper described "a new, up-to-date, moderate priced runabout, very desirable for doctor or city salesman, which I will sell on payments."[24]

New car financing became possible with the advent of sales finance companies in 1913. These companies acted as middlemen: banks supplied capital to the finance companies, finance companies supplied credit to dealers, and dealers then offered credit to their customers. Similar arrangements were already in use to finance expensive consumer goods like pianos. The first large company to enter the auto finance business was the Guarantee Securities Company, which began working with Willys-Overland dealers in 1915. Demand was so great that Guarantee began offering credit for all makes of cars in 1916.[25]

Edsel Ford was fully aware of these developments. In 1916 Edward Rumely, an independent financial consultant, wrote the younger Ford a long memo noting that auto production would soon outpace the number of available cash buyers and that it was necessary to broaden the market. He recommended that the Ford Motor Company set up a separate, company-controlled banking operation for financing new car purchases. But Henry Ford was opposed to consumer credit. He didn't borrow money himself, and he was not about to encourage his customers to do so. Rumely's memo was buried in Edsel's office file.[26]

The financiers who ran General Motors lacked Henry Ford's phobias regarding debt. In 1919 they established the General Motors Acceptance Corporation, just the sort of financing arm Edward Rumely had suggested. GM customers financed 33 percent of their new car purchases in 1919. By 1923 the figure had risen to 46 percent. Ford managers watched glumly as people bought the more expensive Chevrolet because they could pay on time.[27]

Of course, some Ford customers *did* buy their cars on time, because individual dealers arranged credit through local finance companies. But GMAC had greater visibility and public confidence. In 1928, a dozen years after Edward Rumely first recommended it, the Ford Motor Company finally set up its own finance company.[28]

Interestingly, the Ford Motor Company itself seemed not to recognize its greatest marketing coup for what it was. On January 5, 1914, Henry Ford and James Couzens held a news conference to announce the new five-dollar day wage policy. Apparently thinking the story was of strictly local interest, they invited only Detroit newspapers. But by the next day virtually every daily newspaper in the country carried the story, showering the company with effusive praise. In the two weeks following the announcement, the papers in New York City alone filled fifty-two front-page columns with lines like "a bolt out of the blue sky flashing its way across the continent and far beyond" (*Sun*), "an epoch in the world's industrial history" (*Herald*), and "a magnificent act of generosity" (*Evening Post*). Closer to home, the Algonac (Michigan) *Courier* called Henry Ford "one of God's noblemen," while a Detroit *Free Press* headline trumpeted "New Industrial Era Is Marked by Ford's Shares to Laborers." *Ford Times* editor Charles Brownell estimated that the five-dollar day produced over two million lines of glowing newspaper coverage for Henry, his company, and his car.[29]

The man on his way to becoming America's richest citizen suddenly found himself lionized by leftists and labor leaders. The Illinois Federation of Labor proclaimed that the sort of labor-management cooperation represented by Ford's plan "will be the solution of the labor wars in the country." Clarence Darrow said that all American employers must eventually follow Ford's approach: "Labor will demand its own. Ford has recognized the right of labor to make this demand." The Michigan Socialist Party published an admiring pamphlet called *The Bombshell That Henry Ford Fired*.[30]

There was inevitable opposition to the new wage policy. The *Wall Street Journal* railed that Ford "has in his social endeavor committed economic blunders, if not crimes. They may return to plague him and the industry he represents as well as organized society." A corporate official in New York characterized the five-dollar day as "the most foolish thing ever attempted in the industrialized world," an innovation that would "only result in unrest among the laboring classes." And J. J. Cole of Cole Motor Co. summed up the reaction of Ford's fellow automakers: "If Ford wants to amuse himself, all right. He can afford it. Others can't." But opposition by capitalists and businessmen only reinforced the public's high opinion of Ford.

As in the Selden patent case, average citizens saw Henry as the maverick, looking out for the little guy.[31]

Ford may have initially underestimated the powerful symbolism of the five-dollar day, but he quickly recognized it as the golden marketing opportunity it was and moved to capitalize on it. At the New York auto show in January reporters mobbed Ford, and he polished his image by dropping lines like, "I think it is a disgrace to die rich" and "Goodwill is about the only fact there is in life. With it a man can do and win almost anything. Without it he is practically powerless." Over the next few months Ford parlayed a series of interviews, statements, and personal appearances into celebrity status. Historian David Lewis summed up the effect of the five-dollar day: "The year 1914 marks a decided turning point in the public relations status of Henry Ford and the Ford Motor Company. Before that year Ford was known only as one of the nation's more prominent industrialists. From 1914 until his death in 1947, he was easily the world's best-known manufacturer. His company, similarly, enjoyed a considerable reputation prior to 1914; however, from there on it was the best-known business enterprise on the face of the globe."[32]

In early 1917 the Ford Motor Company decided to take advantage of all the free publicity it was generating; it halted national advertising. For nearly six and a half years, Ford did no advertising, arguing that it was passing the reduced overhead on to the customer. Ford got away with this scheme because the Model T sold so well and because the company knew that individual dealers, fearful of having their territories cut if they didn't perform, would advertise on their own. Indeed, total dealer spending on advertising averaged some three million dollars a year during the 1917–1923 period. Finally, after a conference with branch managers, Ford decided to resume national advertising in 1923. Seven million dollars was to be expended yearly on billboards, magazines, and newspapers.[33]

But the automobile market had changed greatly since 1917, and Ford had not kept pace. The Model T was now well behind the technological curve. Other car makers had offered improvements like electric starters, battery ignition, and electric headlights, but these did not become optional on Fords until 1919 and standard until 1926. Cooling systems with water pumps were now as reliable as Ford's simpler thermo-syphon system, and were much less likely to boil over when the engine was working hard. Improved three-speed sliding gear transmissions made the Model T's two-speed planetary transmission no longer defensible. Urban buyers in particular complained about the planetary gearbox because it required them to keep one foot firmly on the low-speed pedal when driving slowly around town. Closed bodies, once only 10 percent of the market, claimed 34 percent in 1923 and 56 percent by 1925. But when Ford put a closed sedan body on a Model T chassis in place of an open touring car body, it transformed a light, nimble 1,200-pound car into a sluggish, top-heavy 1,950-pound car. Improved roads meant that the Model T's high ground clearance, short wheelbase, and flexible three-point suspension were less advantageous. Indeed, on smooth roads the Ford system subjected passengers to a choppy, swaying ride.[34]

But much more than technological change was at work. Ford had ignored the realities of the emerging consumer society. As people's economic circumstances improved, they became accustomed to things they had once regarded as luxuries—running water, electricity, phonographs—and they raised their horizons accordingly. What customers wanted (indeed, what it seems customers always want) was *more*.[35]

When the Model T was young, it did offer more: more flexibility and privacy than the train or the trolley; more speed and comfort than the bicycle or the buggy; and more sheer value than any other automobile in the world. The Model T established a baseline for affordable, reliable, convenient personal transportation. But hardly had that baseline been established than other forces began to raise it. Advancing automotive technology was part of the story, but changing definitions of what constituted "more" were at least as important.

"More" not only meant speed, comfort, and convenience. It also meant prestige, style, and appearance. When the Model T first appeared, ownership of *any* automobile conferred prestige. But over time the automobile market developed its own hierarchy, so that certain cars carried more cachet than others. The cheap, abundant, utilitarian Model T fell to the bottom of that hierarchy. The expansion of the used car market exacerbated that fall. In 1917 there was one car registered for every 21.8 people in the country. By 1923 the ratio had declined to one car for every 8.4 people. This meant there were large numbers of recently purchased upmarket used cars competing in price with new Model Ts.[36]

By 1923 style and appearance were important factors in the sale of new cars. In part this was due to the growing influence of women, who often were the final arbiters in family car buying decisions and who cared more about style and color than most men. But auto executives also realized that as the car market approached saturation, they must give potential customers a reason to sell a two- or three-year-old car and buy a new one. One way to do that was to change styling yearly so

The 1925 Chevrolet challenged Ford with style: two-tone paint, a nickel-plated radiator, and disc wheels—at a reasonable price.

that the old car actually looked old. As early as 1921, General Motors' policymakers acknowledged "the very great importance of styling in selling," so GM made annual model changes every year after 1923. In late 1924, General Motors introduced a new Chevrolet with a nickel-plated radiator, smart-looking solid disk wheels, and lower, longer, smoother styling than the Model T.[37]

Ford's advertising campaigns after 1923 reflect a company unsure how to deal with a changing world. Much of the copy is defensive, saying, in effect, "See, the Model T is actually better than you think it is." One Ford ad asserts that the new closed Fords "offer you not only economical and dependable transportation, but also a more attractive style and greater share of motoring convenience," while another touts, "Deeply cushioned seats, improved interior arrangement, and cowl ventilator. . . .

In 1926 Ford was trying to make the best of outdated features like thermo-syphon cooling.

Wide doors that open forward, revolving type window lifters, enlarged rear compartment and a recess shelf for parcels." These are all features Ford was forced to introduce in 1923 to meet the competition.[38]

In 1926 Ford tried to deal directly with growing criticism of antiquated features like the flywheel magneto, the planetary transmission, and thermo-syphon cooling by issuing a set of advertisements extolling the virtues of those very features. Since Ford ads from 1908 make many of the same arguments, the 1926 material comes off merely as putting the best face on outdated technology.[39]

Having successfully targeted women in the past, Ford tried again in 1925 with an elegant booklet entitled *Her Personal Car*. The new publication features stylishly dressed women playing cards, shopping at upscale stores, and getting into Model Ts parked in front of columned mansions. "In freeing herself from the restraint of the chauffeur-driven car," says one caption, "the fashionable woman of athletic tastes has taken eagerly to the

89 / Selling the Model T

*In freeing herself from the restraint of the chauffeur-driven car, the fashionable woman of athletic tastes has taken eagerly to the good-looking, comfortable Ford Coupe.*

Like jeans at the opera, the Model T looks out of place in this 1925 brochure, *Her Personal Car,* aimed at upscale women drivers.

good-looking, comfortable Ford Coupe." But in truth, the tall, boxy Model T looks out of place against these elegant backgrounds. One can't help but ask why a family that could afford the lifestyle depicted in the brochure would not buy something more expensive and stylish than a Ford.[40]

Beginning in June 1924, a series of artfully composed ads played on Ford's image as the big company that looks out for the little guy. "A Giant Who Works For You," says one; "Servant of the Millions," says another. But the results were discouraging, and the campaign ended in September. Perhaps people had trouble believing that the world's largest automobile company was still the friend of the underdog.[41]

In addition to a resumption of advertising, 1923 finally saw Ford attempt to deal with the growing popularity of buying cars on credit. It launched the Ford Weekly Purchase Plan with great fanfare in April. At first glance the new plan sounded like a credit plan, but it was really a savings plan. Customers

Ford launched the Weekly Purchase Plan with great fanfare, but it was not a success.

made payments of five dollars a week to a bank until they accumulated the price of a new Model T. But a customer could do just as well by opening a savings account. In the first eighteen months, some 400,000 people signed up, but only 131,000 actually completed the program and bought cars. Enrollment dropped off steadily thereafter.[42]

Throughout the Model T's history, Ford always had one surefire marketing tool—lowering prices. By the end of 1920 a touring car could be had for $440, nearly half what the same body style had cost in 1909. If the customer wanted the added convenience of electric starting and demountable rims, the price was still only $535. As competition intensified throughout the 1920s, Ford responded with six price cuts between 1921 and

Ford's last set of price cuts couldn't halt the sales decline. Most of the millions who had the means to buy a Model T also had the means to buy a better-looking, better-equipped, better-performing Chevrolet.

1925. The final cut in December 1924 brought the touring car price to $290 ($375 with starter and demountable rims).[43]

But this was the end of the line. Profits on car sales were down to two dollars *per car*, and made up less than 5 percent of the Ford Motor Company's total profit. The remaining 95 percent came from the sale of by-products (mostly from the expanding River Rouge Plant), interest on bank balances and securities, freight charges on shipments of cars and parts, and the sale of repair parts. There was no room for any more price cuts.[44]

The brutal truth was that the Model T's terminal decline began in 1921, when Ford's share of the U.S. market peaked at nearly 60 percent. Absolute sales peaked in 1923 with over 1.8 million cars, but declined every year thereafter, no matter what the company did. In 1926 a wholesale restyling brought all-steel bodies, a lowered chassis, nickel-plated radiator, and the option of colors on the closed cars. Sales fell by 275,000 cars.⁴⁵

With one notable exception, virtually everyone associated with the Model T in the 1920s realized that its time had passed. Customers compared the Ford to newer designs and took their business to other automakers. Dealers felt the pinch of declining sales and responded by complaining to the home office and by dropping their Ford franchises in favor of more progressive firms. Within the company, engineer Laurence Sheldrick reported that Edsel Ford and Ernest Kanzler openly advocated replacing the venerable "T," while Charles Sorensen, P. E. Martin, Sheldrick himself, and fellow engineers Joe Galamb and Gene Farkas all knew it was time for the Model T to go, even if they were afraid to say so.⁴⁶

The great dissenter was, of course, Henry Ford. For a while he blamed receding sales on the dealers themselves, citing inefficiency, indolence, and lack of the proper "mental attitude." As late as December of 1926 he declared that "the Ford car is a tried and proved product that requires no tinkering." Responding to reports that he would build a new 6-cylinder car, he acknowledged experiments with such machines, but said, "They keep our engineers busy—prevent them from tinkering too much with the Ford car."⁴⁷

But, however reluctantly, Ford did listen to a few voices other than his own. Edsel Ford tactfully but steadily pressed for change. Frank Hadas, whose job included periodic inspection trips to the branch plants, told Henry that "the Chevrolet is making tremendous gains. Its dealers are disparaging the Ford, and especially the planetary transmission." Edward Rumely, who earlier had advocated installment sales, wrote Ford in June 1926 describing a personal experience with a Manhattan crowd that regarded a Model T with pity and disparaged its outdated appearance. Gradually, even Henry Ford admitted that the automobile market had changed. His "Universal Car" was no longer universally loved and desired. On May 26, 1927, the Ford Motor Company announced the end of Model T production. The company would build a new model.⁴⁸

What explains Henry Ford's blindness in the face of years of mounting evidence that the Model T's run was over? One answer lies in the Ford Motor Company's transformation into an absolute monarchy. In 1919 Henry bought out all of his minority stockholders; as sole owner of Ford Motor Company, he could do as he wished. His vision had created one of the world's great industrial enterprises, had created a product that transformed the nation, he reasoned. Why should he listen to the opinions of lesser men? People who challenged Ford found life within the company increasingly difficult, and they began to leave. "Watch the Ford officials go by," joked one Detroit magazine. Joining the stream of departures were sales manager Norval Hawkins and production expert William Knudsen. Both moved to General Motors and became important factors in the rise of Chevrolet. Ernest Kanzler, Edsel's chief lieutenant, wrote a courageous memo outlining why the Model T should be replaced. Within six months he was gone. Those who remained found it prudent to agree with Henry, even when they believed he was wrong. Only Edsel, who wasn't going to leave and wasn't going to be fired, consistently pushed for replacing the Model T.[49]

If we view the Model T at least in part as a work of art, then we have another explanation for Ford's reluctance to give up on it. The car was a deeply personal expression of Ford's aesthetic sense, an aesthetic sense that changed little throughout his life. Ford was not like a jazz musician, always searching restlessly for the next new sound. He was more like a fiddler who plays traditional music within a well-defined genre. He loved the Model T for its inner beauty, and casting that beauty aside was almost too painful to contemplate.

Finally, there is a significant moral dimension to Ford's attitude toward the Model T. Despite his great personal wealth, Ford was genuinely worried about extravagance and high living. Historian Steven Watts has noted that Ford's adamant opposition to tobacco and alcohol reflected his Victorian conviction that "the ability to regulate one's appetites was the key to virtuous character." Henry Ford's disdain for his son's Grosse Pointe friends revolved around his disapproval of activities he regarded as wasteful and decadent, such as golfing, yachting, speedboat racing, drinking, smoking, and vacationing in Europe. When he instituted the five-dollar day, he also instituted a Sociological Department charged with making sure workers spent their newfound wealth responsibly and constructively. By the 1920s Ford was becoming concerned about the corrosive effects of the burgeoning consumer culture. In 1921 he publicly contradicted the principles espoused by his former sales manager Norval Hawkins, telling a reporter, "Advertising? Absolutely necessary to introduce good, useful things; bad when it's used to create an unnatural demand for useless things, as it too often is."[50]

Ford was also uncomfortable with the idea of planned obsolescence and annual model changes based on looks. In his 1923 book *My Life and Work,* written in collaboration with Samuel Crowther, Ford wrote, "It is considered good manufacturing practice, and not bad ethics, occasionally to change designs so that old models will become obsolete and new ones will have to be bought either because repair parts for the old cannot be had, or because the new model offers a new sales argument which can be used to persuade a customer to scrap what he has and buy something new.... Our principle of business is precisely to the contrary. We cannot conceive how to serve the customer unless we make him something that, as far as we can provide, will last forever." These attitudes lead to the conclusion that Ford saw the tough, utilitarian, slowly evolving Model T as not only all the car people would ever need—it was all the car they should ever want. More power, more colors, more style, more comfort were extravagances that simply fed insatiable appetites. A responsible manufacturer didn't encourage such extravagance, just as it did not encourage borrowing money.[51]

It was, of course, a fruitless stand. Ford helped unleash the very attitudes and appetites he now deplored, and they were now beyond his control. For his company to survive, it must accommodate itself to the new market realities. If Henry Ford tried to "stand athwart history, yelling Stop," he would be flattened by a phalanx of brightly colored, nickel-plated, new Chevrolets.[52]

# 5 OWNING & DRIVING THE MODEL T

Driving a modern car with an automatic transmission, cruise and climate control, stereo, and soft, cushy seats is akin to aiming a compact living room down the highway. Compared with driving an early car like a Model T, it is a detached, disconnected experience. To appreciate the difference, let's step over ninety years back in time and take a drive in a 1914 Ford.[1]

Let's assume that a 1914 touring car is parked in front of our house. For convenience, let's assume further that some kind soul has already started the engine for us. As we approach the car, we find it shaking rhythmically in time to the idling motor. The whole car vibrates, not violently but steadily, and the exhaust makes a regular *chuffa-chuffa-chuffa* sound.

When we reach the driver's side, to our consternation there is no door, only the outline of a door embossed into the body. Walking around to the other side of the car, we open the passenger door, step up to the running board, step up again to the floor of the car and settle into the front seat. Sliding across to the driver's position, we can now see why there is no driver's

A 1914 Model T touring car, ready for a drive.

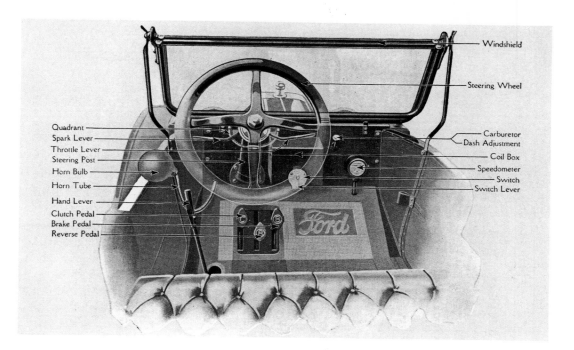

The view from the front seat, as depicted in a 1914 owner's manual. The key controls are the Hand Lever on the left; the Clutch, Brake, and Reverse Pedals on the floor; and the Spark Lever and Throttle Lever under the steering wheel on either side of the steering post.

door. Poking through a slot in the floorboards on the left side of the car is a large lever, looking like it would be more at home in a steam locomotive than in an automobile. If there were a door on that side, the lever would make entering the car too difficult to bother with.

Once actually in place behind the wheel, settled into a seat some three and a half feet above the road, the driver has a commanding view. Model T drivers were generally not wealthy, but they could look down their noses at virtually everyone else on the road. As modern drivers, we hunt in vain for some sort of seat adjustment. In 1914 drivers were expected to adjust themselves to their cars, not the other way around. The only instrument is a speedometer; there is not even a fuel gauge. A large box with a switch on the front is screwed to the vertical dash. It houses the ignition coils; we will come back to it later.

Some things do look familiar, but the front seat of a Model T is an *Alice in Wonderland* world for drivers of the twenty-first century—nothing is what it seems. Take for instance the levers that sprout from either side of the steering column. They remind us of the control stalks found on modern cars that operate turn signals and windshield wipers. But the 1914 Model T has neither of these amenities. The lever on the right is the throttle, or in modern terms the accelerator. It moves through a ninety degree arc along a notched quadrant. Pull it back toward the bottom of the quadrant and the engine speeds up. Push it toward the top of the quadrant and the engine slows down.

The lever on the left is the spark control and also has a ninety degree quadrant. The spark lever determines at what point in the piston stroke the fuel is ignited. That function is automated on modern cars, but don't worry about it now. We will get back to it when we talk about starting the car.

Now we turn to the floor of the car, where three pedals accompany that locomotive-style lever. Anyone familiar with a standard transmission would assume that they are a clutch, a brake, and an accelerator. That assumption would be wrong. The pedal on the right is not an accelerator but rather the brake. It acts directly on the transmission, not the wheels. The middle pedal is for reverse gear. Leave it alone and the car can be driven forward. Hold it down and the car goes backward.

The pedal on the left is the one we will use the most. It operates the clutch, low gear and high gear. Push the pedal down all the way and you are in low gear. Leave it up and you are in high gear. Push it halfway down and you are in neutral, with the clutch disengaged.

With all this in mind, we are finally ready to go for a drive. Put your left foot against the clutch pedal, grasp that locomotive lever with your left hand, and slide it forward. Feel the pedal begin to push backward against your foot? That's because the lever was holding the pedal in the neutral position, with the clutch disengaged. The lever was also engaging the emergency brakes on the rear wheels. In fact, if the car is parked on an incline it will start to roll forward or backward—unless the lever is engaged. But let's assume we're on flat ground. Ease your left foot down, and with your right hand move the throttle lever toward you. In a Model T noise precedes motion. From under the hood comes a rumble as the engine accelerates. From under the floor comes a rising mechanical groan as the low gear band takes hold. Then the car moves forward, either smoothly or with a jerk, depending on how well you coordinate your right hand and left foot.

We are moving, but if you try to go faster by opening the throttle, you find that noise increases while velocity does not. If speed kills, you can live forever in low gear. Low is for climbing steep hills, negotiating swampy mud, pulling stumps, and getting started. We are moving at about seven miles per hour; it is time to shift into high. Ease off with your left foot and push the throttle forward so the engine doesn't race when you pass through neutral. Keep pulling your foot back, pull the throttle toward you again, and the car will settle into high, either smoothly or roughly depending on your skill. Not quite like slipping the lever into "D" and pressing the gas pedal, is it?

Now that we are moving, we must turn our attention to

steering. The first thing you notice is the vibration. The steering wheel and column both shake in sympathy with the engine and the road; your hands feel every bump and rut. Since there is no power steering, the wheel feels rather heavy, especially at low speeds, but when you turn it things happen quickly. The steering requires only one and a quarter turns lock-to-lock, meaning that turning the wheel sharply turns the car sharply. In fact, it is easy to turn too sharply, so one soon learns to turn the wheel only a little and see what happens before adding more steering input.

Turning a corner vigorously can be disconcerting, because a car this high off the ground feels very tippy; it does not inspire confidence. (There is a reason that racing cars are built low and wide, not high and narrow.) Vigorous cornering may also bring with it the smell of gasoline. The fuel tank is under the seat; a bit of sloshing gas escapes through the small vent hole in the gas cap. Such sloshing indicates a nearly full tank, a useful bit of information in a car without a gas gauge.

Driving a Model T or any other open car of this era is a visceral experience. The noise is constant: the engine and transmission whine, the body rattles, the top flaps, the wind blows. You are always aware of your environment: temperature, precipitation, road dust, and outdoor smells. You are also aware of every hill, every bump, every change in the road's surface. You must actively drive this car, paying attention to engine sounds, changes in the rhythm of the rattles, and the adjustment of the hand controls. You must also plan ahead. For instance, spying a hill down the road calls for a decision. If the hill is long or steep, you might have to shift back into low gear. But you may simply be able to open the throttle and accelerate so as to build enough momentum to carry you up and over the hill.

Now that we have been driving for a while, lets see what it takes to stop the car. The Model T has two sets of brakes, neither of them particularly effective. The emergency brakes, actuated by the hand lever and acting only on the rear wheels, are primarily for keeping the car from moving when it is at rest. The transmission brake, actuated by the right foot pedal, is charged with stopping the moving car, but it is barely up to the task. As with hills, the key phrase to remember when stopping a Model T is "plan ahead." Anticipate where you want to stop, push the throttle forward to slow the engine down, apply the brake with your right foot, and when the engine starts to labor, push the left pedal halfway down to disengage the clutch. Gradually the car winds to a stop. "Panic stops" in a Model T live up to their name, usually resulting in a collision with the thing you were trying to avoid.

While we are stopped, let's talk about the middle floor pedal, which engages the reverse. Pull the hand lever back to the neutral position; then push down on the middle pedal with one foot and the car will begin to back up. One of the great tricks available to Model T drivers was the ability to rock the car back and forth when trying to extricate it from mud or deep ruts. With the right foot on the reverse pedal and the left foot on the clutch pedal, the car can easily be shifted back and forth between low and reverse, while the right hand is free to move the throttle as needed.

Thus far we have avoided the ritual of hand-cranking a Model T. That is because a thorough knowledge of the operation of the various controls is essential to start the car safely. A wrong move will make starting difficult at best, hazardous to your health at worst. Starting begins in the front seat and goes like this:

1. Pull the hand lever as far back as it will go to set the emergency brakes and disengage the clutch.
2. Push the spark lever to the top of its quadrant. This fully "retards" the spark, insuring that the fuel/air mixture will not ignite early. This is extremely important, because early ignition will cause the crank handle to kick back, an unfortunate happenstance that can result in a broken wrist.
3. Pull the throttle lever down to the fifth or sixth notch from the top of its quadrant.
4. Flip the switch lever in the center of the coil box (remember that?) all the way left to "magneto."
5. Get out of the car and go around to the front. Now the fun begins. One of the first things an old Model T hand will tell you is not to grab the crank handle with your full fist. Rather, grasp the handle with your four fingers and lay your thumb parallel to the handle. If the engine kicks back, you will want to let go as quickly as possible, so too firm a grip is hazardous.
6. Shove the handle forward to engage the crank ratchet and pull upwards in a quick, firm, clockwise movement. (Never push down on the crank. It is another good way to get hurt if the engine kicks back.) If all goes well the engine will pop to life, but you will still have more to do.
7. Walk smartly to the driver's side, reach into the cabin (there is no door, remember), pull the spark lever down until the engine smoothes out, and push the throttle lever forward until the engine slows down to idle. Now you are ready to go around to the working door, climb in, and go for a drive.

That is a bare-bones description that assumes that everything goes well. In the real world, things often don't go well. For instance, if you have failed to properly set the hand lever, you may think that your Model T is related to Stephen King's malevolent Plymouth, Christine. An improperly set lever leaves the clutch slightly engaged and the rear brakes not fully engaged. The result is a car that creeps forward, threatening to run over the unsuspecting driver. Many a Model T driver has found himself caught between his over-eager car and the wall of his garage or shed, yelling for someone to come and give the hand lever a yank.[2]

Starting your Model T in cold weather can be much more difficult, and Ford made sure that its owner's manuals provided specific instructions for that situation. Inventive owners came up with their own methods. Some used a blowtorch to heat the intake manifold, hoping to vaporize the gasoline more easily. Others drained the water from the cooling system and replaced it with hot water heated on a stove, thus pre-heating the whole engine block. Still others built a fire under the oil pan to warm the engine oil. Some simply swore a lot, which did little to help start the car but made the owners feel better. An electric self-starter finally arrived as optional equipment in 1919, seven years after Cadillac introduced it and well after most other carmakers adopted it.[3]

Driving the Model T after dark or in inclement weather posed special problems.

Until 1915 all Model T exterior lights depended on actual flames. The single taillight and the two side lamps on either side of the windshield burned kerosene, while headlights burned acetylene. To generate the acetylene, cars of the day carried their own miniature chemical plants. A typical acetylene generator had two chambers: an upper one filled with water and a lower one filled with calcium carbide. When water was dripped on calcium carbide, the resulting chemical reaction liberated acetylene gas, which was piped to the headlights through rubber hoses. Lighting the headlights meant turning on the water and then opening the hinged lens on each headlight to ignite the escaping gas with a match. Acetylene gave a bright, white light that was quite effective. In 1915 Ford introduced electric headlights powered by the magneto. While more convenient than the gas lamps, their brightness depended on car speed. The faster the car went, the faster the flywheel magneto turned, and the brighter the light shone. A brisk run down a long hill risked burning out the headlights, while an idling Model T produced a very weak light indeed. It wasn't until 1919, after the introduction of a modern electrical system that included

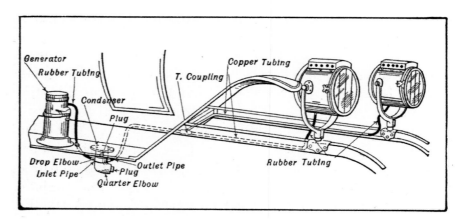

A typical acetylene gas headlight system, from the 1926 edition of *The Model T Ford Car*, by Victor Pagé.

starter, generator, and battery, that Model T owners could have bright, constant headlights independent of engine speed.[4]

Rain, snow, or cold weather were never fun for the Model T owner. Touring cars and roadsters lacked any sort of side glass. Rather, they had side curtains, featuring pieces of clear isinglass mounted in canvas. These were snapped into place between the top and the body, but they served primarily to keep out the big raindrops and large wind gusts while considerably reducing visibility. There were, of course, no heaters, so in cold weather the Model T driver dressed as if he were walking. Winter driving in a touring car was so uncomfortable that many owners simply put their car on blocks or jack stands until spring. It was not until the advent of inexpensive closed bodies in the 1920s that motoring became a year-round activity. Even then, the lack of amenities is startling to modern drivers. It was not until 1925 that Ford offered a single, *hand-operated* windshield wiper as standard equipment. Driving a 1925 Model T in the rain kept the driver busy indeed.[5]

Even checking the fuel level and filling the tank was not simple in the Model T. Remember that the tank was under the front seat, and there was no gas gauge. To check the fuel level you got out of the car, removed the seat cushion, unscrewed the gas cap and stuck a calibrated stick into the tank. Such sticks could be purchased, but they were sometimes given away as free promotional items. The owner could also make one by following instructions found in the owner's manual.

Checking the oil level was no more convenient. It required reaching under the car to open two petcocks on the crankcase. If oil came out of the top petcock, the level was too high. If oil didn't come out of the bottom petcock, the level was too low. If oil came out the bottom but not the top, the level was correct. Some drivers used pliers to keep their hands clean, while others

This factory photo shows a Model T touring car body with its full set of cumbersome side curtains installed.

made or purchased a special long-handled wrench that made the task somewhat easier.

By modern standards, all cars of the early twentieth century required a high level of routine maintenance. Every Model T owner's manual featured an illustration showing all the points on the chassis that needed lubrication, some as often as every 50 miles. Manuals also recommend checking and tightening all nuts and bolts every thirty days, removing and cleaning the muffler "occasionally," and adjusting the transmission bands regularly. The manuals are filled with cutaway drawings showing the internal workings of engine, transmission, and rear axle as well as instructions for more demanding work like removing carbon from the combustion chamber and grinding the valves (both operations requiring removing the cylinder head), setting the valve timing, and disassembling the rear axle.

Some of the advice offered seems quaint to the modern reader, such as the suggestion that engine cooling can be improved by twisting the fan blades to a greater angle and the injunction against putting corn meal, bran, or similar substances in the radiator in an attempt to stop leaks.

In addition to manuals from the factory, Model T drivers had a wide variety of sources for advice on operating and maintaining their cars. The most common was other drivers. As the number of Model Ts rose into the millions, a typical owner could always find another owner with an opinion on how to solve a problem. Whether that opinion was of any value, how-

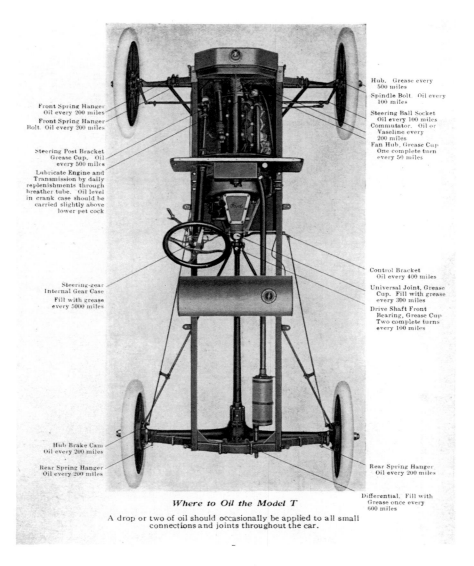

**Where to Oil the Model T**
A drop or two of oil should occasionally be applied to all small connections and joints throughout the car.

If the Model T owner had any spare time, there was always some part on the car that needed lubricating, as this 1909 owner's manual illustrates.

ever, was another matter. More authoritative was the work of Victor Pagé, a prolific author of technical manuals on automotive and aeronautical subjects. Pagé's *The Model T Ford Car*, subtitled "a complete practical treatise explaining the operating principles of all parts of the Ford automobile, with complete instructions for driving and maintenance," went through nine editions between 1915 and 1928. Further help arrived in April 1914 with the first issue of *The Fordowner*, "an independent monthly publication for owners of Ford cars"; it was completely unaffiliated with the Ford Motor Company. Each issue was filled with articles like "Touring With the Ford," "Carburetor Conundrums," "Conserving the Resale Value," and "A Practical Ford Garage."

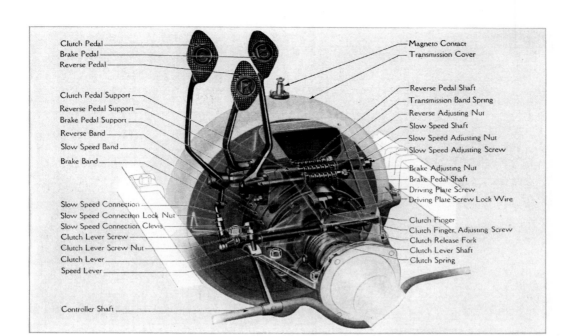

*(top)* Phantom view of transmission, from a Model T owner's manual.

*(bottom)* Grinding valves was once part of the periodic maintenance ritual. This 1914 manual describes how to use a special grinding tool.

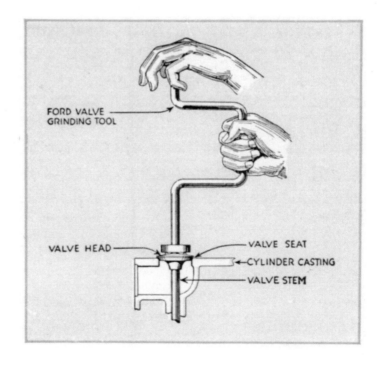

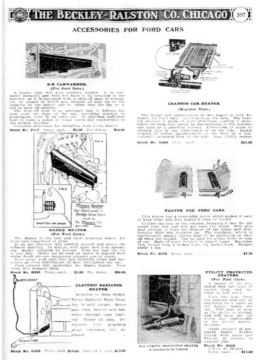

*(left)* This page from a 1922 catalog illustrates a variety of gadgets designed to make checking the Model T's oil and gasoline levels easier.

*(right)* Another catalog page filled with devices intended to make winter driving bearable, if not comfortable.

*The Fordowner* was also filled with advertisements for a vast array of accessories intended to enhance the Universal Car. The Model T may have epitomized standardized production, but it inspired almost unlimited customization. In their memoir *Farewell to Model T*, E. B. White and R. L. Strout noted that "when you bought a Ford, you figured you had a start—a vibrant, spirited framework to which could be screwed an almost limitless assortment of decorative and functional hardware." Some five thousand such gadgets were developed, and many of them actually worked.[6]

Among the most numerous accessories were devices that promised to make cranking a Model T easy and safe; springs and shock absorbers to smooth the car's ride; heaters to make winter driving bearable; gas and oil level gauges to make checking and refilling less onerous; noise-makers like horns and exhaust whistles; "anti-rattlers" that promised to mute the various vibrations endemic to Model Ts; windshield wipers; and complete bodies that could transform Henry's humble machine into a racer or a limousine. A selection of ads and catalog descriptions for some of those five thousand accessories follows.

Model T owners soon learned that their new mechanical wonder was more than mere transportation. It was quickly (top) Nearly every issue of The Fordowner featured several ads for mechanisms claiming to take the effort and risk out of cranking a Model T.

(bottom) Speedster bodies were popular modifications. The author's father learned to drive in a Model T equipped with such a body.

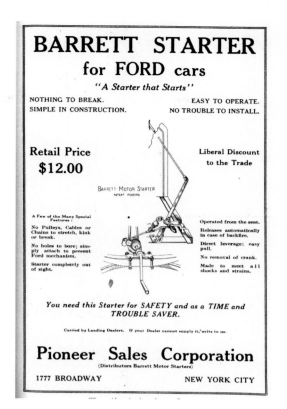

adapted as a portable power source for saw mills, corn shellers, and other farm equipment. Aftermarket equipment was available to turn the car into an effective tractor. And special bodies could always be added to customize a "T" for specific purposes.

## WHAT WILL *YOU* DO *THIS* WINTER?

YOU CAN PUT YOUR FORD IN STORAGE
*and* WALK

YOU CAN DRIVE YOUR FORD AS IT IS
*and* FREEZE

OR

### GET A COZY CAB TOP AND *BE HAPPY*

*Don't put up the car.* You can't get your money out of it if you don't keep it going. Walking is bad in winter. Cold and wet feet spell doctor bills. It's cheaper to ride.

*Don't drive the car without a storm-proof top.* The folding top will keep the sun off—that's all. Loose, flapping curtains won't keep you dry nor warm. They are all right in an emergency—not all winter long.

**DON'T TRY TO GET ALONG WITHOUT A COZY CAB TOP.** It is adjustable, storm-proof, detachable. It does not weigh much, not any more than the folding-top, and it is much more durable. You don't know and don't pay any attention to sleet, rain, snow and zero winds when you are under a COZY CAB Top.

The cost is not much. *Cozy Cab* Top for Roadster is $60.00, for the Touring Car $85.00. Add your saving of doctor bills, livery hire, street car fare and cold cures for one winter, and the top is paid for. Besides, you will have the pleasure of going where you will, when you want to, regardless of the weather, and of getting there quicker.

Write for the illustrated descriptive folder.
IF YOUR DEALER CANNOT SUPPLY YOU WE WILL SHIP DIRECT.

### FOUTS & HUNTER COMPANY
14 Third Street     Established 1873     Terre Haute, Ind.

---

The makers of the Cozy Cab promised to enable Model T owners to use their expensive investments all year round.

*(top)* Ford Times was full of stories about Model T owners using their cars for something other than transportation. Here a flivver drives a saw mill.

*(bottom)* In this Ford Times photo, a farmer uses his Model T to provide power for filling his silo. Note the water hose connected to the radiator in an attempt to keep the engine cool.

*(top)* A special conversion kit with large spiked wheels allows this Model T to pull a reaper.

*(bottom)* The October 1922 issue of *Fordson Farmer* magazine featured an enterprising minister who turned a Model T into a mobile church.

# 6 THE MEANING OF THE MODEL T

Historian David Hounshell called the Model T "the only revolutionary automobile of the twentieth century," because what it did could be done only once. "Its design and mass production made people want an automobile."[1] In 1999 a group of automotive historians and journalists participated in an elaborate process to name "the car of the century." They nominated one hundred cars, winnowed the list down to twenty-five, and revealed the winner at a Las Vegas event in December. The whole effort was anticlimactic, however, because the winner could have been predicted at the start. The humble Model T ranked well ahead of its nearest competitors: the space-efficient Austin Mini Minor, the engineering tour-de-force Citroën DS, the all-time sales champion Volkswagen Beetle, and the exotic Porsche 911. Even though a majority of the survey participants came from Europe, they recognized the singular nature of the Ford and its accomplishment.[2]

The Model T's meaning to American and world history was so profound and multifaceted that it is easy to lose sight of what it meant to the individuals who owned it. The people who bought, drove, and rode in Model Ts lacked the long-term perspective of historians and journalists, but they recognized that the Ford was at the center of something extraordinary. As one resident of Muncie, Indiana, told sociologists Helen and Robert Lynd, "Why on earth do you need to study what's changing in this country? I can tell you what's happening in just four letters: A-U-T-O!" The most common and influential "A-U-T-O" was Henry Ford's Model T.[3]

People like this Muncie resident tried in various ways to come to grips with the changes wrought by the Model T. A common approach was to relate the car to the horse it replaced. In one of the most famous pieces ever written about the Model T, E. B. White and Richard L. Strout repeatedly resorted to horse metaphors. Describing the way the Model T's transmis-

A few of the many humorous Model T post cards.

sion often caused the car to inch forward, even when in neutral, they recalled, "In this respect it was like a horse, rolling the bit in its tongue, and country people brought to it the same technique they used with draft animals." Recounting what happened when the driver crank-started a Model T without fully setting the hand brake, they wrote, "I can still feel my old Ford nuzzling me at the curb, as though looking for an apple in my pocket." Model Ts were subject to various mechanical ailments, but often they seemed to just fix themselves: "Farmers soon discovered this, and it fitted nicely with their draft-horse philosophy: 'Let 'er cool off and she'll snap into it again.'" On another page they describe a helpful ferryboat captain "resetting the bones" of a broken Model T rear axle.[4]

Many owners named their cars, just as they named their horses, but Fords in general came to be referred to by two particular names—*flivver* and *Tin Lizzie*. The genesis of both is uncertain. *The Oxford English Dictionary* calls *flivver* "obscure" and defines it as "a cheap motor car or aeroplane. Also 'a destroyer of 750 tons or less.'" The *New Dictionary of American Slang* also

says that *flivver* is "a Model T Ford car" and "any car, airplane, or other vehicle, esp. a small or cheap one." Both references also indicate a connection between *flivver* and *failure,* which might imply that the name was applied to Model Ts as an ironic joke. The term may have been associated with Fords as early as 1910, but it seems to have come into wide use in the 1920s.[5]

The *OED* defines *Tin Lizzie* simply as "a motor car, esp. an early model of a 'Ford'" and indicates that the phrase first appeared between 1910 and 1920. The *New Dictionary of American Slang* suggests a Southern origin for *Tin Lizzie.* It gives one definition as "any old, ramshackle car or truck," explaining that "such vehicles were sturdy, dependable, and black, like the traditional ideal Southern servant, called *Elizabeth.*" Regardless of how these two terms came about, what they have in common is a friendly, non-threatening feel. Referring to the Model T in this way emphasized the positive, personal aspects of the machine and downplayed the revolutionary changes it was causing.

The foregoing discussion suggests that humor was an important part of people's relationship to the Model T. Humor is one way to deal with new, powerful phenomena. Automobile jokes appeared at the very end of the nineteenth century, and they soon became part of the standard vaudeville repertoire. In fact, Ford jokes antedated the Model T. The company's preference for the low-priced end of the market caused some competitors to disparage its products. "It's not a car, it's just a Ford," was a typical jibe credited to rival salesmen. But when the Model T became ubiquitous across the country, "Ford joke" became synonymous with "Model T joke."[6]

Model T owners themselves provided much of the humor. Some jokes referred to the car's reliability: A man wanted to be buried with his Ford, because it had pulled him out of every hole he had ever been in. A Ford doesn't need gasoline, because it can run on its reputation. What always keeps moving? Ford cars and the earth.[7]

More typical were self-deprecating jokes based on the car's size, price, or idiosyncrasies: What department at the hardware store carried tires for your Ford? The rubber band department. What did the chicken say after being run over by the Ford? "Cheep, cheep, cheep!" When the young man asked his date how she liked riding in his Ford, she said that she was "having a rattling good time."[8] A whole genre of humorous postcards was devoted to Ford jokes, as were a plethora of cheap booklets.

The heyday of the Ford joke was 1914 to 1920, when the production and sales of the car was rising nearly every year and its effects were rapidly spreading across the American landscape. But in the early 1920s, as the public's attitude toward the Ford shifted, so did its attitude toward Ford jokes. Cracks like "A Ford is like a bathtub—you don't like to be seen in one" seemed less like humor and more like the simple truth. When the Keith-Albee vaudeville organization decided that Ford jokes were no longer funny and banned them from its theaters, it was a sign that the automobile market had matured and that the public's love affair with the Model T was ending.[9]

Music also played a role in making sense of the Ford's impact. The archives of the Benson Ford Research Center in Dearborn contain examples of twenty-one songs and instrumental pieces about or inspired by the Model T. Probably the best known was 1914's "The Little Ford Rambled Right Along," in which the driver of a reliable Ford steals the girl from the driver of an unreliable "big limousine." The song emphasized the Model T's ability to level life's playing field, putting rich and poor on an equal footing. "On the Old Back Seat of the Henry Ford," on the other hand, reminded people that the Model T offered

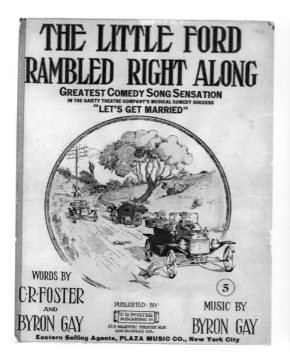 

Two of the most popular Ford songs.

courting couples an unprecedented degree of privacy and mobility.[10]

Much more serious was "Flivver Ten Million," a fantasia written by composer Frederick Shepherd Converse. Featuring actual car horns as well as the sounds of rattles, squeaks, and crashes, the fourteen-minute piece began with the production of the ten millionth Model T and followed it on a cross country trip that included a necking party, a joy ride, and a crash. The program note said that after the accident, the car "proceeds on its way with redoubled energy, typical of the indomitable spirit of America." Despite that optimistic ending, the piece implies a recognition that the Model T had affected people's ideas about morality, leisure time, and acceptable risk. The Boston Symphony Orchestra and the New York Philharmonic both performed the composition to favorable reviews.[11]

Many Model T owners and observers expressed their reaction to the car in poetry. Much of it was clever doggerel, like parodies of the Psalm 23 or "Twinkle, Twinkle, Little Star":

> The Ford is my auto; I shall not want another.
> It maketh me to lie down beneath it; it soureth my soul.

or

> Beat it, beat it, little car
> How I wonder what you are;

115 / The Meaning of the Model T

Climbing up the hills on high;
Passing all the others by.[12]

Other verses showed considerable insight. In 1917 a "Brief History of the U.S." reduced the American story to five words, presaging the comment made to the Lynds in Muncie:

Columbus
Washington
Lincoln
Roosevelt
Ford[13]

And the February 1915 issue of *Ford Owner* printed the following rather startling piece:

*I Am the Ford!*

I am the Ford.
I lend wings to the feet of all men. I am the magic carpet of the multitude.
I devour distance. I am the dream come true.
Where man is, there am I.
Him I carry over deserts in equatorial heat; over ice in Arctic cold I carry him.
I assist in evil deeds in the black night, and carry help and healing within the hour.
In war, I am the agent of destruction—I bring death.
In war, I move on errands of mercy—I bring life.
In peace, I widen the horizon—I cause the world to shrink.
I work—and I lighten the toil of the worker.
I play—and stir the joy of living in millions.
I am the liberator—I loose the iron grasp of the cities.
I bring health and hope into spent lives.
To the pale toiler of the crowd, I show the sunset from the high hills, and the chains slacken.
I show the warm meadows at noon to the household drudge—the spiritless childbearer. She raises her eyes, and the yoke lightens.
I revive old longings. Travel I make possible, and the seeing of strange things.
I take men upon my shoulders and carry them to the end of the rainbow.
Over gray lives I cast the golden haze of possibility—romance—and life becomes bearable.
I bring youth, and the high hopes of youth.
I am the forerunner of a new age.
I am the Ford.[14]

Here are many of Norval Hawkins's "ideas about goods"—magic carpets, shrinking worlds, slackening chains, high hopes of youth. These are the dreams Ford owners believed their cars could fulfill. But the author also notes the car's use in crime and in war and recognizes full well that the Ford brings something more than personal fulfillment. Ford drivers are active participants in the coming of a new age.

But the new age heralded by the Ford was the most dynamic and fast-paced era in American history. The Model T was a catalyst for changes that made the Model T itself obsolete. As roads improved, incomes rose, and tastes changed, the sturdy, idiosyncratic Model T lost its charm. Newer "magic carpets" offered bright colors, easier operation, higher speeds, more comfort, more style.

Henry Ford acknowledged these changes only reluctantly, but finally even he admitted that the Model T's time was over. When the last flivver came down the line at Highland Park on May 26, 1927, most Americans recognized that history had turned a page. A few, like a well-to-do lady in Montclair, New Jersey, rejected the future. She purchased seven new Model Ts and stored them away so that she could drive one for the rest of her life. Others wrote heartfelt letters to Henry Ford describing what the Model T meant to them. A farmer from Kapowsin, Washington, told of a used Ford roadster that served him for thirteen years without an overhaul or any other major maintenance. "I do not know how many thousands of miles it has run," he related, "as the speedometer was worn out when I bought it and I never put another one on, but it has been in constant use." The *Roanoke News* summed up the reaction of many: "It will be long before America loses its affectionate, if somewhat apologetic, remembrance of the car that first put us on wheels. We probably wouldn't admit it to anyone, but deep in our hearts we love every rattle in its body."[15]

Of course, the end of Model T production was not the end of the Model T. In 1927, 11,325,521 of them were still registered in the United States. By 1931 attrition had cut the number to 5,432,000. When R. L. Polk and Company attempted the last actual count in 1948, they found 73,111 cars still registered, with an untold number resting unregistered in barns and sheds. One result of this longevity was that the Ford became the subject of oral tradition. Former owners passed stories about road trips, crank-starting, and mastering the mysteries of three-pedal control down to children who never saw an actual Model T. Because more stories were told about Fords than any other brand, the general public gradually came to regard all "old cars" built before 1930 as Model Ts.[16]

Another important transmitter of Model T lore was a 1936 essay first published in *The New Yorker* under the title "Farewell, My Lovely." Using the pseudonym "Lee Strout White," E. B. White and Richard L. Strout wrote what amounted to an epitaph for the Model T. They described the Ford as "the miracle God had wrought. And it was patently the sort of thing that could only happen once. Mechanically uncanny, it was like nothing that had ever come to the world before. Flourishing industries rose and fell with it. As a vehicle, it was hard-working, commonplace, heroic; and it often seemed to transmit those qualities to the persons who rode in it." Published as a book under the title *Farewell to Model T* and reprinted in several anthologies, the essay's engaging prose kept the Tin Lizzie alive for those who could not remember the car itself. This author first encountered it in high school about 1960.[17]

But the Model T was not relegated to mere folklore. After World War II the car experienced a rebirth as an object of specialist and enthusiast interest. The Model T became a "collectible." In his book *The Shape of Time*, art historian George Kubler noted that "the retention of old things has always been a central ritual in human societies." The modern impulse to put things in museums, he said, is related to the ancestor cults of primitive tribes. Both aim "to keep present some record of the power and knowledge of vanished peoples." Kubler also observed that things that are merely useful—mixing bowls, lawn sprinklers, Crescent wrenches—are more likely to be discarded than things "made for emotional experience." The Model T's creators may have conceived it as merely useful, but Norval Hawkins and Ford's customers understood that the Model T and all other automobiles were also made for emotional experience.[18]

Rebuilding and driving obsolete automobiles as a hobby began in the 1930s, but it really blossomed after World War II. Behind this renewed interest in old cars were the sorts of emotional factors found in Hawkins and Kubler's ideas. They included:

- Nostalgia for the past: many people who owned and drove old cars wanted to relive past experiences.
- Fulfillment of unfulfilled wishes: some people wanted to own these cars when the cars were new but were too young or too poor to do so.
- Desire for a more visceral driving experience: the advent of all-steel bodies, heaters, radios, and plush seats in the 1930s and power steering and automatic transmissions in the 1950s disconnected drivers from the highway.

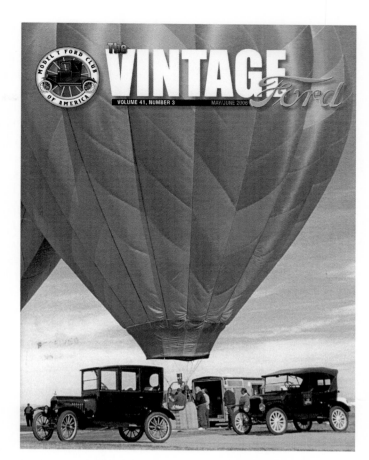

The two major Model T Ford clubs publish high-quality bimonthly magazines for Model T collectors.

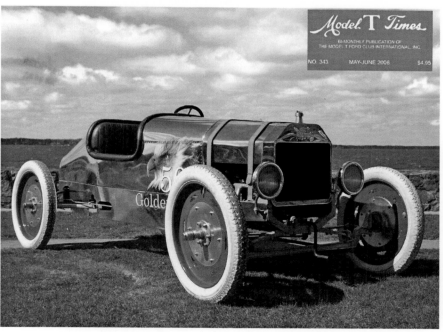

119 / The Meaning of the Model T

Antique cars offered a more basic, engaging driving experience.
- Desire to work on something mechanical: old cars were more technically accessible and forgiving than new cars.[19]

More Model Ts survived than any other make, so they attracted more collectors and hobbyists than any other make. Fords were available, cheap, and easy to work on—characteristics that had made them popular when they were new as well. In 1952 enthusiasts formed the Model T Ford Club, later renamed the Model T Ford Club International. Members restored their cars to "like new" condition, drove them on tours, gathered for shows, and judged the best restorations. In 1966 the Model T Ford Club of America was organized. Today both clubs thrive, with some twelve thousand members between them.

The growth of the Model T as a hobby spawned a substantial cottage industry in reproduction parts, accessories, and services. A sampling of the two club magazines (*Model T Times* and *Vintage Ford*) revealed advertisements for over forty companies catering to the needs of Tin Lizzie collectors. The publications also advertised dozens of tours and shows sponsored by the two organizations. Both clubs recognize that members with direct memories of the flivver's heyday are dwindling, and both actively recruit and offer activities aimed at younger people. They work hard to keep the Model T alive.

A very different approach to keeping the Model T alive is taken by hot rodders. Almost from its inception, the Model T was modified to make it go faster. Special cylinder heads, camshafts, pistons, and suspension parts allowed owners to improve the performance of their cars. Similar "speed parts" were made for the Ford Model A and V8 cars of the 1930s. Cars so equipped raced on oval tracks across the country, but in southern California the proximity of huge dry lake beds east of Los Angeles gave people a place to drive at high speeds for several miles. Most cars running on the dry lakes were modified Ford roadsters, *hot roadsters* in the local parlance. The term was eventually shortened to *hot rod*.[20]

Like antique car collecting, hot rodding blossomed after World War II. Although initially concentrated in southern California, the sport spread across the country. The typical hot rod was a pre-war Ford roadster with a Ford flathead V8 engine. More modern overhead valve engines came into vogue as the 1950s progressed. Model Ts were often chosen as the starting point for a hot rod, and builders exhibited enormous creativity in building machines that reflected their own aesthetic

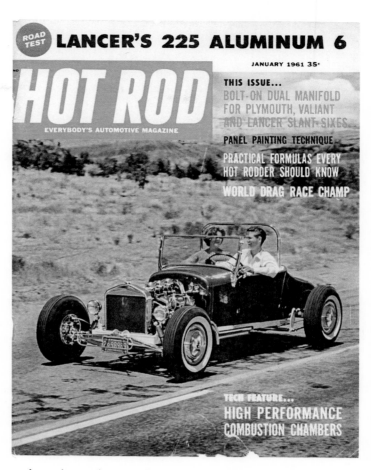

Magazines such as *Hot Rod* regularly featured highly modified Model Ts.

and mechanical tastes. While the antique car club members blanched at the "destruction" of perfectly restorable Model Ts by hot rodders, the two groups had much in common. Both enjoyed actually working on their vehicles, and both relished the visceral driving experience offered by the cars they had restored or created. The hot-rodded Model T roadster, with a big, chromed engine and flames painted on the body, joined the perfectly restored Tin Lizzie in the general public's mind as representing the Model T.

As significant as the Model T was at the micro level, it was nevertheless its larger, macro meaning that caused it to be labeled the car of the century.

To work for the Ford Motor Company between 1907 and 1914 was to be part of one of the most astonishingly creative periods in the history of any American corporation. In those years Ford designed and built a revolutionary automobile, designed and built a revolutionary factory, created within that factory a revolutionary production system, and capped it all with a revolutionary change in the relationship between work and wages. In the process it also profoundly shaped the direction the new century would take. As one observer put it decades later, "Henry Ford's Highland Park Plant was the place where the mainspring of the twentieth century was wound."[21]

Two major elements of the century go hand in hand: mass production and mass consumption. Henry Ford (actually his ghost-writer William Cameron) defined mass production as "the focusing upon a manufacturing project the principles of power, accuracy, economy, system, continuity, speed, and repetition. The normal result is a productive organization that delivers in continuous quantities a useful commodity of standard material, workmanship and design at minimum cost. The necessary precedent condition of mass production is capacity, latent or developed, of mass consumption, the ability to absorb large production. The two go together."[22]

Both the methods of mass production and the sales methods necessary to promote mass consumption were spawned and perfected in the auto industry, with Ford leading the way. That the Ford Motor Company eventually lost its lead to General Motors in no way diminishes the Ford accomplishment. Producers of other consumer goods like refrigerators, washing machines, vacuum cleaners, and radios quickly adopted mass production methods. The American standard of living came to mean the purchase, discard, and re-purchase of large quantities of machine-made goods. To be sure, not all industries were equally amenable to mass production. Efforts to apply it fully to housing and to furniture met with very mixed success.[23] On the other hand, elements of mass production made their way into unexpected places. On "factory farms" thousands of standardized pigs or chickens are raised in large buildings, fattened by a mechanically fed stream of standardized fodder. McDonald's and Taco Bell deliver "in continuous quantities" food of "standard material, workmanship and design at minimum cost."

When World War II broke out, American mass production industries made a remarkably quick conversion to producing war material. None of the other belligerents could match the ability of the United States in turning out guns, helmets, tanks, ammunition, and combat boots. Assembly line techniques were even adapted, with considerable difficulty, to aircraft and ships. American aircraft factories more than kept up with the appalling losses over Europe, while American shipyards built Liberty and Victory ships faster than two Axis navies could sink them.

Ford's five-dollar day is often cited as a key factor in expanding the middle class. But less often understood is just how that happened. The five-dollar day did more than simply increase wages. It reversed the historical relationship between wages and skill. Throughout history, the way for workers to increase the price they demanded for their services was to increase their skill level. The master craftsman always made more money than the journeyman. Conversely, the way for an employer to lower labor costs was to lower the skill required to do the work. For example, mechanization in the textile industry and the shoe industry lowered the skill level required to spin yarn and make shoes, and lowered the value of the labor of the workers in question. But the five-dollar day turned that relationship on its head by creating something the world had never seen before: the low-skill/high-wage job. Suddenly high-wage jobs were available to large numbers of people who could never have had them before, especially people from rural areas and from foreign countries. The Georgia sharecropper and the Polish peasant both found in Detroit or other industrial cities the opportunity to make a good living despite their lack of industrial skills. Unfortunately, in cities like Detroit, this process led to a devaluing of education on the part of many workers and their children. Why do I need an education, they asked, to work on the line? A willingness to work, not a high school diploma, is all that is required.[24]

But it turned out that the reversal of the wage/skill relationship was not permanent. As the automobile industry fought to meet competition from foreign cars in the 1980s, especially from Japan, it became clear that the Japanese had a different approach to assembly line work. Japanese automakers had discovered that they could increase quality and productivity by actually involving their workers in the improvement of the process. They did not want workers who parked their brains at the time clock and picked them up again at the end of the shift. They wanted workers who were educated, engaged, and who could do a variety of different jobs in the manufacturing process. Not only did they want their employees to have a high school education, they even wanted them to have training beyond high school. To compete in quality and productivity, the U.S. industry gradually adopted the same approach. The consequences for anyone seeking an entry-level job at an auto plant were profound. Those workers who had concluded that education was not necessary for affluence now found that the rules had changed. No longer was the industry an "opportunity gate," opening wide to anyone willing to work hard. It became instead an "opportunity turnstile," open only to those with sufficient education and skill.[25]

The five-dollar day had a major unintended consequence. Henry Ford's new wage policy constituted an unwritten contract with his workers: they submitted to the discipline of the assembly line, and he paid unprecedentedly high wages. But the large mass of unskilled workers thus created also had a great deal of latent power. If they chose to withhold their labor, they could bring a company or an industry to a halt. That power lay dormant until the Great Depression broke that unwritten contract. When Ford and other assembly line–based industries could no longer pay the high wages, workers were no longer willing to submit to work discipline. They responded by joining industrial unions based not on craft skills but on common employment in an industry. Making their latent power manifest in the form of sitdown strikes, they ultimately forced employers to sign written contracts defining a new balance of power between worker and employer. Thus it was that Henry Ford, who hated labor unions, unwittingly created the conditions that gave rise to an organized labor movement that would remain a potent social and political force for the remainder of the century and beyond.[26]

The most visible legacy of the Model T was mass automobility. Before the Model T, auto ownership was restricted to the wealthy and the upper middle class. The Model T placed auto ownership within the reach of far more people, and it planted the desire for auto ownership in the mind of nearly everyone. The consequences of this shift are too massive to summarize in a short essay, but a few of the most salient are worth noting.

The auto industry became the driving force in the twentieth-century American economy, and the steel, oil, and rubber industries grew rich fulfilling its needs. Highway construction, virtually insignificant at the beginning of the century, grew steadily, fed by gasoline taxes willingly paid by drivers who wanted better roads. The culmination of this building boom was the Interstate Highway System, one of the great public works projects in human history, on the scale of China's Great Wall or Rome's aqueducts.

The American propensity for owning a house in the middle of a piece of land, no matter how small, created "streetcar suburbs" in the nineteenth century. But mass automobility facilitated the growth of vast new suburbs, with their attendant schools, retail stores, and industries. The depopulation of older cities like Detroit and Buffalo and the expanding population of cities like Houston and Atlanta would have been impossible without ready access to automobiles.

Mass ownership of cars not only has allowed us to drastically alter our landscape, it also has drastically altered our atmosphere. Once viewed as preferable to the manure and urine deposited by horses, auto exhaust gradually came to be understood as a serious problem. Successful efforts to reduce emissions from individual cars are vitiated by increases in the sheer numbers of cars and in the miles people drive. The long-term atmospheric consequences of the twentieth-century choice of mass automobility are hotly debated in the twenty-first century, as are the possible solutions.

Mass automobility is something to die for—literally. Since the late 1930s traffic fatalities have averaged between forty and fifty thousand people per year. Great strides have been made in making both cars and roads safer, but as with air pollution, these efforts are offset by the increase in the number of miles driven. Deaths per mile have fallen steadily, but rising mileage keeps total annual deaths about the same.

Finally, in the early years of the twenty-first century, it is difficult to think about automobiles without also thinking about oil, its price, its availability, its location. The huge Spindletop strike near Beaumont, Texas, in 1901 meant that the automobile boom would be fueled by abundant and cheap gasoline. Americans came to view that condition as the natural order of things, even after domestic wells could no longer meet domestic demand. But it turns out that some of the world's most abundant oil fields are in some of the world's most politically volatile places—or perhaps those places are volatile because they contain abundant oil fields. Maintaining a high level of automobility meant becoming deeply involved with those places, for better or worse.

By the time the last Model T was produced on May 26, 1927, it was obsolescent technology, superseded by its more powerful, more comfortable competitors. But the Model T had nothing to apologize for. It had done enough.

## NOTES

Epigraph: Henry Ford, in collaboration with Samuel Crowther, *My Life and Work* (Garden City, NY: Garden City Publishing Company, 1922), 73. The statement appears to have first been published in the Ford Motor Company publication *Ford Times,* June 1913, 366.

### Introduction

1. In his 1922 autobiography, *My Life and Work,* written in collaboration with Samuel Crowther, Henry Ford claims to have told his staff in 1909 that "any customer can have a car painted any color that he wants so long as it is black." For more on Model T colors, see chapter 3.

### Chapter 1. Automobility in 1908

1. The Duryeas' claim as the first American automobile *manufacturers* is well established. The question of who built the first American automobile is much murkier. The Duryeas often get credit, but Richard Scharchburg has demonstrated that several individuals preceded them. Settling the question of who was first, if possible at all, is far beyond the scope of this book. See Richard P. Scharchburg, *Carriages without Horses: J. Frank Duryea and the Birth of the American Automobile Industry* (Warrendale, PA: Society of Automotive Engineers, 1993).

2. Automobile Manufacturers Association, Inc., *Automobiles of America* (Detroit: Wayne State University Press, 1962), 104.

3. "Eighth Annual Review of Complete Motor Cars," *Cycle and Automobile Trade Journal,* March 1, 1908, 45–132. Despite the magazine's claim that its listing was all-inclusive, there were so many small manufacturers making only a few cars that a truly comprehensive listing was virtually impossible to compile. The 166 makers of gasoline, steam, and electric cars listed in *Cycle and Automobile Trade Journal* were certainly the major manufacturers in America at the time.

4. Peter Hugill, "Technology and Geography in the Emergence of the American Automobile Industry, 1895–1915," in *Roadside America: The Automobile in Design and Culture,* ed. Jan Jennings

(Ames: Iowa State University Press, 1990), 31. The French company Panhard and Levassor pioneered front-engined autos in 1891, hence the designation "French-type." Mercedes pushed the design's possibilities farther than Panhard did.

5. James J. Flink, *America Adopts the Automobile, 1895–1910* (Cambridge, MA: MIT Press, 1970), 210; Hugill, "Technology and Geography," 33; and Christopher Wells, "Car Country: Automobiles, Roads, and the Shaping of the American Landscape, 1890–1929," Ph.D. dissertation, University of Wisconsin–Madison, 2004, 50–51.

6. Hugill, "Technology and Geography," 34–36; and Wells, "Car Country," 56–58.

7. Wells, "Car Country," 57–59.

8. Wells, "Car Country," 22.

9. Flink, *America Adopts the Automobile*, 64–68; and Clay McShane, *Down the Asphalt Path: The Automobile and the American City* (New York: Columbia University Press, 1994), 176–79.

10. Flink, *America Adopts the Automobile*, 70.

11. The terms *horse-minded* and *transportation-minded* come from Wells, "Car Country," 30.

12. Wells, "Car Country," 32; and Flink, *America Adopts the Automobile*, 92.

13. Wells, "Car Country," 34–37.

14. Wells, "Car Country," 23; and Flink, *America Adopts the Automobile*, 100, 101–2.

15. Wells, "Car Country," 37; and Flink, *America Adopts the Automobile*, 102.

16. Wells, "Car Country," 51–52; and Flink, *America Adopts the Automobile*, 235–36.

17. Wells, "Car Country," 59.

### Chapter 2. Creating the Model T

1. Allan Nevins, *Ford: The Times, the Man, the Company* (New York: Charles Scribner's Sons, 1954), 150–237; and Sidney Olsen, *Young Henry Ford* (Detroit: Wayne State University Press, 1997), 60–189.

2. Nevins, *Ford: The Times, the Man, the Company*, 274–83, 323–41.

3. Henry Ford, in collaboration with Samuel Crowther, *My Life and Work* (Garden City, NY: Garden City Publishing Company, 1922), 24; Nevins, *Ford: The Times, the Man, the Company*, 276; and *The Automobile*, January 11, 1906, 107, 119.

4. Nevins, *Ford: The Times, the Man, the Company*, 324–27.

5. Beverly Rae Kimes and Henry Austin Clark, *Standard Catalog of American Cars, 1805–1942* (Iola, WI: Krause Publications, 1985), 163, 940, 156.

6. Nevins, *Ford: The Times, the Man, the Company*, 47–48, 147–59; and interview with Oliver Barthel, Accession 65, Ford Motor Company Oral History Project, Benson Ford Research Center, The Henry Ford, Dearborn, Michigan, 27.

7. Ford R. Bryan, *Henry's Lieutenants* (Detroit: Wayne State University Press, 1993), 269–90, 123–24, 267–68; and Trent Boggess, "Edward Huff, Henry Ford, and the Flywheel Magneto," *Vintage Ford,* March–April 1996, 22–23.

8. Nevins, *Ford: The Times, the Man, the Company,* 348–49; and *Ford Motor Cars 1907, Models N and S* (Detroit: Ford Motor Company, 1907), 22–23.

9. George Brown, one of the clerks working at the Piquette Plant, described how Edsel would come from school, throw his books on Brown's desk, and disappear into the experimental room: "He was in that experimental room every day after school, and we wouldn't see him when we went home. They'd be in there continuously in that experimental room." Interview with George Brown, Accession 65, Ford Motor Company Oral History Project, Benson Ford Research Center, The Henry Ford, Dearborn, Michigan, 181.

10. Trent Boggess, "Birthplace of the Model T: The Piquette Avenue Plant," *The Vintage Ford,* November–December 1998, 20–21; and interview with Joseph Galamb, Accession 65, Ford Motor Company Oral History Project, Benson Ford Research Center, The Henry Ford, Dearborn, Michigan, 16.

11. Steven Watts, *The People's Tycoon: Henry Ford and the American Century* (New York: Alfred A. Knopf, 2005), 357; and Allan Nevins and Frank Hill, *Ford: Expansion and Challenge: 1915–1932* (New York: Charles Scribner's Sons, 1957), 19–20.

12. Nevins and Hill, *Ford: Expansion and Challenge,* 618–19.

13. Eugene Ferguson, *Engineering and the Mind's Eye* (Cambridge, MA: MIT Press, 1992), 23.

14. Interviews with John Wandersee and Joseph Galamb, Accession 65, Ford Motor Company Oral History Project, Benson Ford Research Center, The Henry Ford, Dearborn, Michigan, 20, 26; and Henry Ford, in collaboration with Samuel Crowther, *My Life and Work,* 69, 53.

15. "Ninth Annual Review of Complete Motor Cars," *Cycle and Automobile Trade Journal,* March 1, 1909, 121, 103, 113.

16. *Horseless Age,* January 13, 1909, 48, and January 6, 1909, 31.

17. P. M. Heldt, *The Gasoline Automobile, Its Design and Construction: Volume I, The Gasoline Motor* (New York: The Horseless Age Co., 1911), 88–89.

18. Christopher Wells, "Car Country: Automobiles, Roads, and the Shaping of the American Landscape, 1890–1929," Ph.D. dissertation, University of Wisconsin–Madison, 2004, 170–71.

19. Interview with John Wandersee, Accession 65, Ford Motor Company Oral History Project, Benson Ford Research Center, The Henry Ford, Dearborn, Michigan, 22–25.

20. Engineers would eventually learn how to build workable planetary transmissions with more than two speeds. Today's ubiquitous automatic transmissions have planetary gear sets offering three, four, or more forward speeds. For a discussion of automotive transmission history, see Philip G. Gott, *Changing Gears: The Devel-*

*opment of the Automotive Transmission* (Warrendale, PA: Society of Automotive Engineers, 1991).

21. Interview with Joseph Galamb, Accession 65, Ford Motor Company Oral History Project, Benson Ford Research Center, The Henry Ford, Dearborn, Michigan, 15.

22. "The Motor Car of Nineteen-Nine," *Motor,* January 1909, 50.

23. *Horseless Age,* February 3, 1909, 175–76.

24. Boggess, "Edward Huff, Henry Ford, and the Flywheel Magneto," 20–33.

25. "The Motor Car of Nineteen-Nine," 49–107.

26. *Ford Motor Cars, Model T, Advance Catalog,* Ford Motor Company, Detroit, Michigan, 1908; Bruce McCalley, "The Developing Model T," and Trent Boggess, "Even More Pre-Production Data," both at the Model T Ford Club of America website, available at www.mtfca.com/encyclo/1908.pdf.

27. Trent Boggess, "Model T #1," *Vintage Ford,* November–December 2004, 17–24.

## Chapter 3. Manufacturing the Model T

1. Allan Nevins, *Ford: The Times, the Man, the Company* (New York: Charles Scribner's Sons, 1954), 648.

2. The principal authors who have examined the Model T production process are Nevins, *Ford: The Times, the Man, the Company*; David A. Hounshell, *From the American System to Mass Production, 1800–1932: The Development of Manufacturing Technology in the United States* (Baltimore: Johns Hopkins University Press, 1984); Stephen Meyer, *The Five Dollar Day: Labor Management and Social Control in the Ford Motor Company, 1908–1921* (Albany: State University of New York Press, 1981); and Lindy Biggs, *The Rational Factory: Architecture, Technology, and Work in America's Age of Mass Production* (Baltimore: Johns Hopkins University Press, 1996). Nevins and Hounshell focus on the transition to the assembly line, while Meyer is most concerned with the workers' experience. Biggs discusses the development of Highland Park after 1914, but her primary concern is the plant's architecture rather than the production process.

3. John M. Staudenmaier, "Two Technocrats, Two Rouges: Henry Ford and Diego Rivera as Contrasting Artists," *Polhem* 10, no. 1 (1992), 14, 11; and Allan Nevins and Frank Hill, *Ford: Expansion and Challenge: 1915–1932* (New York: Charles Scribner's Sons, 1957), 610.

4. For a discussion of Ford's assembly process as typical of the early auto industry, see Nevins, *Ford: The Times, the Man, the Company,* 220–49.

5. Biggs, *Rational Factory,* 90; and interview with Fred Rockleman, Accession 65, Ford Motor Company Oral History Project, Benson Ford Research Center, The Henry Ford, Dearborn, Michigan, 30.

6. Nevins, *Ford: The Times, the Man, the Company,* 281–82.

7. Nevins, *Ford: The Times, the Man, the Company*, 281–82; Hounshell, *From the American System to Mass Production*, 221–23; and Charles E. Sorensen, *My Forty Years with Ford* (Detroit: Wayne State University Press, 2006), 93–96, 45.

8. Nevins, *Ford: The Times, the Man, the Company*, 340; and Hounshell, *From the American System to Mass Production*, 226.

9. Plan of Highland Park redrawn by the author from Accession 73, Highland Park Appraisal, 1919, Box 8, Benson Ford Research Center, The Henry Ford, Dearborn, Michigan; and Horace Arnold and Fay Faurote, *Ford Methods and the Ford Shops* (New York: Engineering Magazine, 1915), 24–25.

10. Arnold and Faurote, *Ford Methods*, 36; and *Factory Facts from Ford*, Accession 951, Ford Motor Company Non-Serial Imprints (Detroit: Ford Motor Company, 1915), 13; available at Benson Ford Research Center, The Henry Ford, Dearborn, Michigan.

11. Fred Colvin, "Building an Automobile Every 40 Seconds," *American Machinist*, May 9, 1913, 757–62; and Nevins, *Ford: The Times, the Man, the Company*, 454n. The original negative of this photograph no longer exists at the Benson Ford Research Center.

12. Hounshell, *From the American System to Mass Production*, 239–41; and interview with William Klann, Accession 65, Ford Motor Company Oral History Project, Benson Ford Research Center, The Henry Ford, Dearborn, Michigan, 21–22.

13. Arnold and Faurote, *Ford Methods*, 74.

14. Arnold and Faurote, *Ford Methods*, 78.

15. Arnold and Faurote, *Ford Methods*, 82.

16. "Factory Has Gigantic Presses," *Ford Times*, December 1913, 133.

17. Fred Colvin, "Ford Radiators and Gasoline Tanks," *American Machinist*, September 4, 1913, 393–96.

18. "Painting Wheels for Model Ts," *Ford Times*, August 1911, 313.

19. Hounshell, *From the American System to Mass Production*, 234–37.

20. Fred Colvin, "Building an Automobile Every 40 Seconds," 757–62; Fred Colvin, "Special Machines for Small Auto Parts," *American Machinist*, September 11, 1913, 439–43; and Hounshell, *From the American System to Mass Production*, 234–37.

21. *Ford Industries*, Accession 951, Ford Motor Company Non-Serial Imprints (Detroit: Ford Motor Company, 1924), 11; available at Benson Ford Research Center, The Henry Ford, Dearborn, Michigan; and Arnold and Faurote, *Ford Methods*, 136.

22. Interview with William Klann, Accession 65, Ford Motor Company Oral History Project, Benson Ford Research Center, The Henry Ford, Dearborn, Michigan, 22–23.

23. Interview with William Klann, Accession 65, Ford Motor Company Oral History Project, Benson Ford Research Center, The Henry Ford, Dearborn, Michigan, 51; and Hounshell, *From the American System to Mass Production*, 246–48.

24. Hounshell, *From the American System to Mass Production*, 251.

25. Hounshell, *From the American System to Mass Production*, 256.

26. Arnold and Faurote, *Ford Methods*, 136–40.

27. Fred Colvin, "Continuous Assembling in Modern Automobile Shops," *American Machinist*, August 26, 1915, 365–70.

28. Fred Colvin, "Continuous Assembling in Modern Automobile Shops," 365–70.

29. Arnold and Faurote, *Ford Methods*, 136–39.

30. Arnold and Faurote, *Ford Methods*, 148–50.

31. *Factory Facts from Ford*, 25.

32. Hounshell, *From the American System to Mass Production*, 256–59.

33. Biggs, *Rational Factory*, 118–24.

34. Trent Boggess, "The Customer Can Have Any Color He Wants—So Long As It's Black," *Vintage Ford*, November–December 1997, 34, 37; Arnold and Faurote, *Ford Methods*, 386–88; and Biggs, *Rational Factory*, 123. The Boggess article is a goldmine of useful information on Ford's painting practices.

35. Arnold and Faurote, *Ford Methods*, 360–61.

36. Boggess, "The Customer Can Have Any Color He Wants—So Long As It's Black," 30.

37. Arnold and Faurote, *Ford Methods*, 360–67; and Boggess, "The Customer Can Have Any Color He Wants—So Long As It's Black," 36.

38. Biggs, *Rational Factory*, 129; and *Ford Industries*, 17, 25, 26, 27.

39. *Factory Facts from Ford*, 61; Gerald T. Bloomfield, "Coils of the Commercial Serpent," in *Roadside America: The Automobile in Design and Culture*, ed. Jan Jennings (Ames: Iowa State University Press, 1990), 42; and Allan Nevins and Frank Hill, *Ford: Expansion and Challenge: 1915–1932* (New York: Charles Scribner's Sons, 1957), 255–57.

40. Nevins, *Ford: The Times, the Man, the Company*, 357–58; and Mira Wilkins and Frank Ernest Hill, *American Business Abroad: Ford on Six Continents* (Detroit: Wayne State University Press, 1964), 26–28.

41. Wilkins and Hill, *American Business Abroad*, 39–50, 434–35.

42. Biggs, *Rational Factory*, 140–60; and *Ford Industries*, 35–61.

43. Hounshell, *From the American System to Mass Production*, 276–79; and Nevins and Hill, *Ford: Expansion and Challenge: 1915–1932*, 431–32.

## Chapter 4. Selling the Model T

1. For a good summary of the design modifications, see Les Henry, "The Ubiquitous Model T," in *Henry's Wonderful Model T, 1908–1927*, ed. Floyd Clymer (New York: McGraw-Hill, 1955). For a detailed examination of all of the changes, see Bruce McCalley,

*Model T: The Car that Changed the World* (Iola, WI: Krause Publications, 1994).

2. Norval A. Hawkins, *The Selling Process: A Handbook of Salesmanship Principles* (Detroit: Norval A. Hawkins, 1920), 68.

3. Steven Watts, *The People's Tycoon: Henry Ford and the American Century* (New York: Alfred A. Knopf, 2005), 94; David Lewis, *The Public Image of Henry Ford: An American Folk Hero and His Company* (Detroit: Wayne State University Press, 1976), 36–37; and Allan Nevins, *Ford: The Times, the Man, the Company* (New York: Charles Scribner's Sons, 1954), 227.

4. Lewis, *Public Image of Henry Ford,* 20–24.

5. Nevins, *Ford: The Times, the Man, the Company,* 299–302; and Lewis, *Public Image of Henry Ford,* 20–24.

6. Accession 19, Ford Motor Company Advertisements, Box 134, Folder 1904; Box 2, Folder 1903–1904; Box 139, Folder 1906–1907; Benson Ford Research Center, The Henry Ford, Dearborn, Michigan.

7. Nevins, *Ford: The Times, the Man, the Company,* 342–43.

8. Nevins, *Ford: The Times, the Man, the Company,* 344.

9. Nevins, *Ford: The Times, the Man, the Company,* 344.

10. Nevins, *Ford: The Times, the Man, the Company,* 343.

11. Nevins, *Ford: The Times, the Man, the Company,* 387–88.

12. Nevins, *Ford: The Times, the Man, the Company,* 396.

13. Hawkins, *Selling Process,* 68.

14. "Why Doesn't More Auto Copy Talk My Language?" *Ford Times,* November 1910, 86.

15. Accession 175, Ford Motor Company Product Literature, *Ford Motor Cars,* Ford Motor Company, Detroit, Michigan, 1911, 5; *Ford Motor Cars,* Ford Motor Company, Detroit, Michigan, 1910, 5; Accession 19, Ford Motor Company Advertisements, Box 2, Folder 1912; and Accession 951, Ford Motor Company Non-Serial Imprints, Box 7, *The Doctor and His Car,* 5; all available at Benson Ford Research Center, The Henry Ford, Dearborn, Michigan.

16. Accession 175, Ford Motor Company Product Literature, *Ford Motor Cars, Souvenir Booklet 1910,* Ford Motor Company, Detroit, Michigan, 1910; *Ford Motor Cars,* Ford Motor Company, Detroit, Michigan, 1911; and Accession 19, Ford Motor Company Advertisements, Box 2, Folder 1913–1914; all available at Benson Ford Research Center, The Henry Ford, Dearborn, Michigan.

17. Accession 175, Ford Motor Company Product Literature, *Ford—The Quality Car,* Ford Motor Company, Detroit, Michigan, 1911; and *Ford Times,* November, 1910, 91; both available at Benson Ford Research Center, The Henry Ford, Dearborn, Michigan.

18. Accession 19, Ford Motor Company Advertisements, Box 2, Folder 1913–1914, Benson Ford Research Center, The Henry Ford, Dearborn, Michigan.

19. Accession 951, Ford Motor Company Non-Serial Imprints, Box 28, *The Lady and Her Motor Car,* Ford Motor Company, Detroit, Michigan, 1911; and Box 55, *The Woman and the Ford,* Ford

Motor Company, Detroit, Michigan, 1916; both available at Benson Ford Research Center, The Henry Ford, Dearborn, Michigan.

20. Lewis, *Public Image of Henry Ford,* 44–45.

21. Lewis, *Public Image of Henry Ford,* 44–45.

22. Charles Hyde, *Riding the Rollercoaster: A History of the Chrysler Corporation* (Detroit: Wayne State University Press, 2005), 221; and Lewis, *Public Image of Henry Ford,* 113–14.

23. Lendol Calder, *Financing the American Dream: A Cultural History of Consumer Credit* (Princeton, NJ: Princeton University Press, 1999), 186.

24. Calder, *Financing the American Dream,* 185.

25. Calder, *Financing the American Dream,* 187–89.

26. Calder, *Financing the American Dream,* 189–91; and Accession 6, Box 1, Folder "Subject File 1913–1916 Automotive Financing, Agricultural Credit Co." April 14, 1916, Benson Ford Research Center, The Henry Ford, Dearborn, Michigan.

27. Calder, *Financing the American Dream,* 192, 195.

28. Calder, *Financing the American Dream,* 195, 199.

29. Lewis, *Public Image of Henry Ford,* 69–71; and Watts, *People's Tycoon,* 179.

30. Watts, *People's Tycoon,* 185, 196, 195.

31. Watts, *People's Tycoon,* 193–94.

32. Watts, *People's Tycoon,* 191; and Lewis, *Public Image of Henry Ford,* 60.

33. Allan Nevins and Frank Hill, *Ford: Expansion and Challenge: 1915–1932* (New York: Charles Scribner's Sons, 1957), 262–63.

34. Nevins and Hill, *Ford: Expansion and Challenge,* 417; Arthur J. Kuhn, *GM Passes Ford: Designing the General Motors Performance Control System* (University Park: Pennsylvania State University Press, 1986), 281; and Beverly Rae Kimes and Henry Austin Clark, *Standard Catalog of American Cars, 1805–1942* (Iola, WI: Krause Publications, 1985), 10, 551–59.

35. Nevins and Hill, *Ford: Expansion and Challenge,* 398.

36. Kuhn, *GM Passes Ford,* 280; and Nevins and Hill, *Ford: Expansion and Challenge,* 394–96. Per capita auto registration figures based on information in Ben J. Wattenberg, ed., *The Statistical History of the United States from Colonial Times to the Present* (New York: Basic Books, 1976), 8, 716.

37. Nevins and Hill, *Ford: Expansion and Challenge,* 399; and Alfred P. Sloan Jr., *My Years with General Motors* (Garden City, NY: Anchor Books, 1972), 310, 192.

38. Accession 19, Ford Motor Company Advertisements, Box 3, Folders 1923 A–F and 1923 G–N, Benson Ford Research Center, The Henry Ford, Dearborn, Michigan.

39. Accession 19, Ford Motor Company Advertisements, Box 5, Folders 1926 D–F, J–M, P–S, and T, Benson Ford Research Center, The Henry Ford, Dearborn, Michigan.

40. Accession 175, Ford Motor Company Product Literature, *Her Personal Car,* Ford Motor Company, Detroit, Michigan, 1925,

Benson Ford Research Center, The Henry Ford, Dearborn, Michigan.

41. Nevins and Hill, *Ford: Expansion and Challenge,* 264; Watts, *People's Tycoon,* 353; and Accession 19, Ford Motor Company Advertisements, Box 139, Folder 1924, Benson Ford Research Center, The Henry Ford, Dearborn, Michigan.

42. Nevins and Hill, *Ford: Expansion and Challenge,* 267–68.

43. Nevins and Hill, *Ford: Expansion and Challenge,* 264.

44. Nevins and Hill, *Ford: Expansion and Challenge,* 266–67.

45. Kuhn, *GM Passes Ford,* 312; Editors of Automobile Quarterly, *The American Car since 1775: The Most Complete Survey of the American Automobile Ever Published* (New York: L. Scott Bailey, 1971), 140; and Nevins and Hill, *Ford: Expansion and Challenge,* 264.

46. Nevins and Hill, *Ford: Expansion and Challenge,* 421–25, 409.

47. Nevins and Hill, *Ford: Expansion and Challenge,* 415–16, 421.

48. Nevins and Hill, *Ford: Expansion and Challenge,* 428–30.

49. Watts, *People's Tycoon,* 289; and Nevins and Hill, *Ford: Expansion and Challenge,* 409–11.

50. Watts, *People's Tycoon,* 309, 363, 352.

51. Nevins and Hill, *Ford: Expansion and Challenge,* 412.

52. I have borrowed the "stand athwart" line from William F. Buckley, who used it in 1955 to describe the point of view of his new magazine, *National Review.* It appeared in the November 19, 1955 issue, on page 5.

### Chapter 5. Owning & Driving the Model T

1. Unless otherwise noted, this chapter is based on 1911–1914 Model T owner's manuals, conversations with Model T enthusiasts David Liepelt and Ken Kennedy, and the author's own limited personal experience at the wheel of a Model T.

2. Floyd Clymer, *Henry's Wonderful Model T, 1908–1927* (New York: McGraw-Hill, 1955), 19.

3. Clymer, *Henry's Wonderful Model T,* 18.

4. Clymer, *Henry's Wonderful Model T,* 35.

5. Clymer, *Henry's Wonderful Model T,* 127.

6. Lee Strout White (pseud.), *Farewell to Model T* (New York: G. P. Putnam's Sons, 1936), 6. Lee Strout White was a pseudonym for E. B. White and Richard L. Strout.

### Chapter 6. The Meaning of the Model T

1. David A. Hounshell, *From the American System to Mass Production, 1800–1932: The Development of Manufacturing Technology in the United States* (Baltimore: Johns Hopkins University Press, 1984), 278.

2. Richard A. Wright, "The Car of the Century? Ford's Model T, of Course," *Detroit News,* December 20, 1999, available at http://info.detnews.com/joyrides/story/index.cfm?id=75.

3. Robert S. Lynd and Helen Merrell Lynd, *Middletown* (New York: Harcourt Brace and World, 1929), 251.

4. Lee Strout White (pseud.), *Farewell to Model T* (New York: G. P. Putnam's Sons, 1936), 3, 17, 25, 28.

5. *Oxford English Dictionary Online*, 2006, available at www.oed.com; and Robert E. Chapman, ed., *New Dictionary of American Slang* (New York: Harper and Row, 1986).

6. David Lewis, *The Public Image of Henry Ford: An American Folk Hero and His Company* (Detroit: Wayne State University Press, 1976), 212.

7. Lewis, *Public Image of Henry Ford*, 123; and General Post Card Collection, Benson Ford Research Center, The Henry Ford, Dearborn, Michigan.

8. Lewis, *Public Image of Henry Ford*, 123; and *Rattling Ford Jokes No. 40* (Baltimore: I & M Ottenheimer, 1916), 5.

9. Lewis, *Public Image of Henry Ford*, 125; and General Post Card Collection, Automobiles—Characters and Cartoons, Box 2, size J, Benson Ford Research Center, The Henry Ford, Dearborn, Michigan.

10. Accession 163, Ford Songs, Box 1, Benson Ford Research Center, The Henry Ford, Dearborn, Michigan; and Floyd Clymer, *Henry's Wonderful Model T, 1908–1927* (New York: McGraw-Hill, 1955), 183–84.

11. Lewis, *Public Image of Henry Ford*, 52–53.

12. *Funny Stories About the Ford* (Hamilton, OH: Presto Publishing Co., 1915), 12, 47.

13. Carleton B. Case, compiler, *Ford Smiles: Jokes About a Rattling Good Car* (Chicago: Shrewesbury Publishing Co., 1917), 7.

14. *The Fordowner*, February 1915, 1.

15. Allan Nevins and Frank Hill, *Ford: Expansion and Challenge: 1915–1932* (New York: Charles Scribner's Sons, 1957), 432–33; and Lewis, *Public Image of Henry Ford*, 193.

16. Lewis, *Public Image of Henry Ford*, 195.

17. Lee Strout White (pseud.), *Farewell to Model T*, 2; and Lewis, *Public Image of Henry Ford*, 194, 517n21.

18. George Kubler, *The Shape of Time: Remarks on the History of Things* (New Haven, CT: Yale University Press, 1962), 79–80.

19. Three major antique car clubs were formed in the 1930s: the Antique Automobile Club of America (1935), the Horseless Carriage Club of America (1937), and the Veteran Motor Car Club of America (1938). The Classic Car Club of America was formed in 1952.

20. Robert C. Post, *High Performance: The Culture and Technology of Drag Racing, 1960–1990* (Baltimore: Johns Hopkins University Press, 1994).

21. The author is indebted to former colleague Scott Dennis for this felicitous metaphor. Given Henry Ford's love of watches and clocks, I think he would have appreciated it as well.

22. Quoted in Melvin Kranzberg and Carroll Pursell, *Technology in Western Civilization, Volume II* (New York: Oxford University Press, 1976), 92.

23. Hounshell, *From the American System to Mass Production*, 311–16.

24. Many authors have examined the changing relationship between wages and skill. Three prominent works on the subject are David S. Landes, *The Unbound Prometheus: Technological Change and Industrial Development in Western Europe from 1750 to the Present* (Cambridge, MA: Cambridge University Press, 1969); Harry Braverman, *Labor and Monopoly Capital: The Degradation of Work in the Twentieth Century* (New York: Monthly Review Press, 1974); and Alan Dawley, *Class and Community: The Industrial Revolution in Lynn* (Cambridge, MA: Harvard University Press, 1976).

25. James M. Rubenstein, *Making and Selling Cars: Innovation and Change in the U.S. Automotive Industry* (Baltimore: Johns Hopkins University Press, 2001), 166–69.

26. Rubenstein, *Making and Selling Cars*, 125–50.

## FURTHER READING

The most comprehensive study of Henry Ford, his company, and his motorcars is the great trilogy written by Allan Nevins and Frank Hill as part of the Ford Motor Company's fiftieth anniversary celebration. The first two volumes, *Ford: The Times, the Man, the Company* and *Ford: Expansion and Challenge: 1915–1932*, recount Henry Ford's early life, the origins of the company, and the rise and fall of the Model T. The third volume, *Ford: Decline and Rebirth: 1933–1962*, takes the company through the Depression, World War II, and the postwar years.

Anyone intimidated by the massive Nevins and Hill study has many other options. Two good introductions to the story of Ford and the Model T are Sidney Olsen's *Young Henry Ford* and Steven Watts's *The People's Tycoon: Henry Ford and the American Century*. Olsen mined the Ford family and business records to create a lively, well-illustrated account of Henry Ford's first forty years, including the two failed auto companies that preceded the Ford Motor Company. Steven Watts's insightful book is perhaps the best single-volume biography of Ford. A third useful book is David Lewis's *The Public Image of Henry Ford: An American Folk Hero and His Company*, which details Ford's transformation from unknown mechanic to near-mythical industrialist. Two books that carry the Ford story past the founder's death and into recent times are Robert Lacey's *Ford: The Men and the Machine* and Douglas Brinkley's *Wheels for the World: Henry Ford, His Country, and a Century of Progress, 1903–2003*. Henry Ford's own autobiography, *My Life and Work*, was ghostwritten by Samuel Crowther in 1922 when Ford was at the peak of his power. It is useful as a window into Ford's mind but should not be taken as a reliable historical account.

Only two of Ford's chief subordinates, Charles Sorensen and Harry Bennett, wrote memoirs. Sorensen's *My Forty Years with Ford* is far superior to Bennett's *We Never Called Him Henry*, but as with most memoirs, both must be taken with a grain of salt. Harry Barnard's *Independent Man: The Life of Senator James Couzens* is an excellent study of the irascible, indispensable man who ran the business side of the Ford Motor Company until 1915. *Henry's Lieutenants*, by Ford R. Bryan, is a useful set of biographical sketches of thirty-eight people who helped Ford create his empire and keep it running. Bryan spent a highly productive retirement exploring aspects of the Ford story others had largely ignored. His *Beyond the Model T: The Other Ventures of Henry Ford* deals with Henry Ford's many interests outside the auto industry; his *Clara: Mrs. Henry Ford* is the best biography of Henry's wife.

The classic work on the Highland Park Plant is *Ford Methods and the Ford Shops*, published in 1915 by Horace Arnold and Fay Faurote. The authors detail the factory's layout and operation just after the assembly line was fully implemented. Fred Colvin's ten articles for *American Machinist* in 1913 cover Highland Park just before the advent of the assembly line. The best synthesis of developments at the great Ford factory is found in David Hounshell's *From the American System to Mass Production, 1800–1932*. Lindy Biggs's *The Rational Factory: Architecture, Technology, and Work in America's Age of Mass Production* places Highland Park in the context of other factories of the time and explores some of the changes made after 1914.

Two books deal directly with the workers' experience at Highland Park, Stephen Meyer's *The Five Dollar Day: Labor, Management, and Social Control in the Ford Motor Company* and Clarence Hooker's *Life in the Shadows of the Crystal Palace, 1910–1927: Ford Workers in the Model T Era*. David Gartman's *Auto Slavery: The Labor Process in the American Automobile Industry, 1897–1950* examines the whole auto industry but of necessity includes extensive coverage of events and conditions at Ford.

The literature on the Model T itself is often aimed at enthusiasts. Two of the most useful for the general reader are *Henry's Wonderful Model T, 1908–1927* by Floyd Clymer and *Tin Lizzie: The Story of the Fabulous Model T Ford* by Philip Van Doren Stern. Both deal extensively with the technical aspects of the car but also explore its social impact. The Clymer book contains a useful essay summarizing the many changes in the "unchanging" Model T. The definitive technical study of the Model T is Bruce McCalley's *The Car that Changed the World*. Extensively illustrated, it includes detailed descriptions of each model year's features. For those interested in the Tin

Lizzie as race car, there is *Model T Ford in Speed and Sport* by Harry Pulfer. Murray Fahnestock's *The Model T. Ford Owner* is a compilation of articles from *Fordowner* magazine. James L. Kenealy's *Model T Ford Authentic Accessories, 1909–1927* illustrates the seemingly endless variety of aftermarket items intended to improve on Henry's original design. Good accounts of life with a Model T include *Coast to Coast in a Model T* by Henry and John Bowerman and *Travels with Zenobia: Paris to Albania by Model T Ford: A Journal* by Rose Wilder Lane. Finally, *Farewell to Model T,* credited to the pseudonymous Lee Strout White (actually E. B. White and Richard L. Strout), provides an evocative coda to the Flivver story.

More general studies of the automobile include *America Adopts the Automobile* and *The Automobile Age,* both by James Flink; *The American Automobile: A Brief History* by John B. Rae; *Down the Asphalt Path: The Automobile and the American City* by Clay McShane; and *Making and Selling Cars: Innovation and Change in the U.S. Automotive Industry* by James Rubenstein. *The Standard Catalog of American Cars, 1806–1942* is a fascinating, monumental effort by Beverly Rae Kimes and Henry Austin Clark to chronicle every pre–World War II American automobile make, no matter how obscure.

# INDEX

Page numbers in *italics* refer to illustrations.

Abell, Oliver J., 33
Accessories for Model T, 106–7, *106–7*
Acetylene gas headlights, 101, *102*
Acme, 83
*Advance Catalog,* 30, 78, *79*
Advertising, 74, 79–81, *80, 82*; after 1923, 88–90, *89*; in club magazines, *119,* 120; to doctors, 81; halting and resumption of, 86; Henry Ford's opinion of, 94; of Model T accessories, 106, *106–7*; outdated technology and, 89, *89*; to women, 82, 89–90, *90*
Aftermarket equipment, 107, *108–10*
Air pollution, 124–25
*American Machinist,* 33, 40, 49
American roads, 3–4, *5,* 23, 31, 83, 87, 124
Anderson, John W., 16
Annual style changes, 87–88, 94–95
Antique car clubs, 118–20
Apple, Vincent G., 28
Arnold, Horace, 33, 53, 54
Artistic vision, 20–21, 33–34
Assembly lines: for chassis, *56–62, 57–62*; for dashboards, 55, *55*; dislike by workmen, 63; effect on production output and cost, 50–51; for flywheels, 53, *53*; Ford's first use of, 53–54; for motors, 54, *54*; time required for Model T assembly on, 60; for war material in World War II, 122
Association of Licensed Automobile Manufacturers, 75
Aufière, Charles Marie François, 28
Austin Mini Minor, 111
Auto emissions, 124–25
Auto industry in American economy, 124

*Automobile, The,* 17, 21
Automobile manufacturers, 1–2
Automobility, 1–12, 124–25

Babcock Electric, *2*
Barthel, Oliver, 18
Battery production, *69,* 70
Bell, David, 18
Bellevue Avenue Plant, 36, *36*
Bishop, Jim, 18
Bodies: closed, 87, 102; colors of, 67; customized, 107, *107–8*; painting of, 65, *65,* 67, *67*; purchase from outside suppliers, 65, *65,* 66; temporary installation of, 62, *62*; varnishing of, 67
*Bombshell That Henry Ford Fired, The,* 85
Brakes, 98, 99
Branch plants, 70–71, *70–71,* 93
Briggs Manufacturing, 66
Brown, George, 77
Brownell, Charles, 85
Brush 1907, 17–18, *18*
Buick, 17, *18,* 21
Burroughs, John, 20

Cameron, William, 122
Carbon, removal from combustion chamber, 103
Cash sales, 83–84
Cato, George, 18
Chassis, 23–25; assembly line for, *56–62, 57–62*; front suspension, 23–24, *23–24*; rear suspension, 24, *24–25*; station assembly of, 50, *50*; temporary body installation on, 62, *62*
Chevrolet, 84, 88, *88,* 93, 94, 95
Chicago meat packing plant, 52, *52*
*Chicago Tribune,* 84

Chrysler Corporation, 73, 83
Citroën DS, 111
Clutch, 98
Cole, J., 85
Cole Motor Company, 85
Collectors of Model Ts, 118–20
Colors of Model T, 67
Columbia and Electric Vehicle Company, 75
Colvin, Fred, 33
Converse, Frederick Shepherd, 115
Cornering, 99
Cost of automobiles: early Ford models, 14; effect of assembly line on production output and, 50–51; electric cars, 2; financing and, 84–85; high-wheelers, 4; Mercedes-type cars, 4; Model N *vs.* comparable cars, 17–18; Model T, 12, 84, 91–92; runabouts, 7; steam cars, 3
Couzens, James, 19, 63, 74, 78, 85
Cozy Cab, 108, *108*
Creation of Model T, 18–31; adjustment to uneven terrain, 23, *26*; chassis, 23–25, *23–25*; detachable cylinder head, 22, *22*; engine, 22; engine mounting, 24–25, *26*; front suspension, 23–24, *23–24*; gasoline consumption, 31; horsepower, 22; left-hand steering, 30, *30*; magneto, 28–29, *28–29*; men involved in, 18–20; prototypes, 20; rear suspension, 24, *24–25*; shakedown trip, 31, *31*; steel, 22–23, 31; transmission, 27, 27–28, 29, 86–87; weight, 21–22, 31
Credit sales, 83–84, 90
Crowther, Samuel, 94
*Cycle and Automobile Trade Journal,* 2, 4, 7, 12, 17
Cylinder block, 42–44, *42–44*

Daimler, 3
Darrow, Clarence, 85
Dashboard: assembly line for, 55, *55*; assembly stands for, 48, *48*, 55; installation of, 59, *59*
Dealers, 74, 77–78, 86, 93
Degener, August, 23
Demand for Model T, 78–79
Design changes, 74, 86, 87–88
Design engineering, 20–21
Detroit Automobile Company, 13, 19
Detroit Edison Electric Illuminating Company, 13
*Detroit Free Press,* 75
*Detroit News,* 75
"Disassembly line" principle, 52
Distribution points for cars, 78

Dodge, John and Horace, 34
Driving as recreation, 10–12, *11*
Driving the Model T, 96–102; after dark, 101–2; checking fuel and oil levels, 102–3; controls for, *97*, 97–98; cornering, 99; on hills, 99; by hobby enthusiasts, 118–20; in inclement weather, 102, *103*; noise from, 98, 99; procedures for moving forward, 98; in reverse, 100; starting the engine, 96, *96*, 100–101; steering, 99; stopping, 99
Duryea, Charles and Frank, 1
Duryea Motor Wagon Company, 1, *1*, 2, 13

Edison, Thomas, 81
Electric cars, 2, 7, 9; Babcock Electric, 2; cost of, 2; disadvantages of, 12
Electric self-starter, 101
Electric Vehicle Company, 75
Employees at Highland Park Plant, 32–33, 64; dislike of assembly line, 63; five-dollar day wages for, 63, 85–86, 122–24; turnover of, 63
Employee unionization, 124
*Engineering Magazine,* 33
Engine of Model T, 22; assembly of, 49, *49*; detachable cylinder head of, 22, *22*; installation of, 58, *58*; mounting of, 24–25, *26*; one-piece cylinder block of, 22; starting of, 61, *61*, 96, *96*, 100–101
"Everyman's Car," 17
Exporting automobiles, 71

*Factory Facts from Ford,* 62
Faraday, Michael, 28
*Farewell to Model T,* 106, 118
Farkas, Gene, 93
Farmers, and early automobile, *8*, 8–9, 11
Farming, Model T as power source for, 107, *109–10*
Faurote, Fay, 33
Ferguson, Eugene, 21
Financing car purchases, 84–85, 90
Five-dollar day wage, 63; media/business reactions to, 85; skill level and, 123; symbolism of, 85, 86
Flanders, Walter, 36, 42
Flink, James, 8
*Flivver,* 112–13
"Flivver Ten Million," 115
Flywheel magneto, 28–29, *28–29*; assembly line for, 53, *53*
Folding top installation, 68, *68*
Folklore about Model T, 117–18
Ford, Edsel, 20, 73, *73*, 84, 93, 94
Ford, Henry, *13*, 14, *73*; artistic

vision, 20, 33–34; creation of Model T, 18–31; interest in large-scale manufacturing, 16; moral principles, 94; *My Life and Work*, 94; opposition to planned obsolescence, 94–95; promotion as great inventor, 80–81; reluctance to end Model T, 93–94, 117; selling idea of, 76–77, *77*; as sole company owner, 93–94; use of five-dollar wage as marketing tool, 86; vision for automobile, 16–17, 18–19, 21
Ford branches and agencies, 74, 77–78
Fordite, 70
Ford Manufacturing Company, 14, 16, 36, 37
*Ford Methods and the Ford Shops,* 33, 64
Ford Motor Company, 14, 16, 19, 31, 37; as absolute monarchy, 93–94; corporate symbol for, 74, 76–77; creation of Model T, 18–31; creativity in 1907–1914 period, 121–25; early models produced by, 14–17, *15, 16*; financing of new car purchases, 84, 85, 90; foreign expansion of, 71–72; manufacturing plants of, 32–73; public relations benefit of five-dollar day wage, 86; Selden patent suit, 75, *76*; selling Model T, 74–95; share of U.S. market in 1921, 92
Ford Motor Company of Canada, 71
*Fordowner, The,* 104, 106
Ford logo, 74, 76–77
*Fordson Farmer,* 110
*Ford Times,* 31, 44, 47, 78, 80, 81, 85, *109*
Ford Weekly Purchase Plan, 90–91, *91*
Foreign expansion, 71–72
Fuel tank, 99, 102; checking fuel level, *102*; installation of, 57, *57*

Galamb, Joseph, 19, 20, 21, 27, 93
Garagiola, Joe, 83
Gas lamps, 101, *102*
Gasoline-powered cars, 2, 3, 7; advantages of, 12; high-wheeler, 4, *6*, 7; Mercedes-type, 3–4, *6*; runabouts, 7, *7*; Selden patent for, 75
General Motors, 73, 88, 94, 122
General Motors Acceptance Corporation, 84–85
Grinding valves, 103, *105*
Guarantee Securities Company, 84

Hadas, Frank, 93
Hand-cranking a Model T, 100–101; accessories for, 106, *107*
*Harper's Weekly,* 10
Hawkins, Norval, 74, 76–78, 80, 82, 94, 117, 118
Haynes-Apperson Company, 2
Headlights, 101–2, *102*
Henry Ford Company, 13
*Her Personal Car,* 89
Highland Park Plant, *32,* 32–33, 34, *37,* 37–51, *53–69,* 53–70, 122; chassis assembly line of, *56,* 57; crane way of, 40, *40,* 65, *65*; cylinder block machining operations of, 42–44, *42–44*; daily car production at, 70; dashboard assembly line of, 55, *55*; dashboard assembly stands of, 48, *48, 55*; design of, 38, *38*; engine assembly at, 49, *49*; five-dollar day wage at, 63, 85–86, 122–24; flywheel assembly line of, *53,* 53–54; foundry mechanization in, 41, *41*; installing folding tops in, 68, *68*; labor turnover at, 63; monorail conveyer of, 38–39, *38–39*; motor assembly line of, 54, *54*; new shop in 1914, 64, *64*; painting car bodies in, 65–67, *65–67*; radiator core assembly at, 46, *46*; stamping transmission covers at, 45, *45*; station assembly of chassis at, 50, *50*; wheel painting at, 47, *47*
Highway construction, 124
High-wheeler, 4, 7; cost of, 4, 9; disadvantages of, 7; Holsman 1908, *6*
Hill, Frank, 20
Hobby enthusiasts, 118–20
Holsman, *6,* 22
Horse-based economy, 8
Horse-drawn commercial vehicles, 9–10, *10*
*Horseless Age,* 22, 28, 76, 81
Horseless carriage design, 3, 4; Selden patent for, 75
Horse-minded auto advocates, 9
Horsepower of Model T, 22
*Hot Rod, 121*
Hot rodding, 120–21, *121*
Hounshell, David, 54, 111
Huff, Ed "Spider," 18, 19, 28–29
Humor, *113,* 113–14

Illinois Federation of Labor, 85
Ingersoll millers, 43, *43*
Internal combustion engine, 25–27
Interstate Highway System, 124
*Iron Age,* 33
Isinglass, 102
Itala, 83

Japanese automakers, 123
John R. Keim Company, 45
Jokes about Model T *113*, 113–14

Kahn, Albert, 38, 42
Kansas City, Missouri, branch plant, 70, *70*
Kanzler, Ernest, 93, 94
Klann, William, 41, 53, 54
Knudsen, William, 94
Kubler, George, 118
Kulick, Frank, 83

Labor unions, 63, 124
*Lady and Her Motor Car, The*, 82
Lead Cab Trust, 9
Left-hand steering, 30, *30*
Lewis, David, 86
Lights on Model T, 101–2, *102*
Lightweight automobile, 12, 83. *See also* Weight of automobiles
Loans for car purchase, 84
Lubrication, 103, *104*
Lynd, Helen and Robert, 111

Mack Avenue Plant, 34, *34*
Magneto, 28–29, *28–29*; assembly line for, 53, *53*
Maintenance of Model T, 103–4, *104–5*
Malcomson, Alexander, 14, 16, 17, 19, 36
Manchester, England, branch plant, 71, *71*, 72
Manufacturing plants, 32–73; Bellevue Avenue Plant, 36, *36*; branch plants, 70–71, *70–71*, 93; Highland Park Plant, *32*, 32–33, 34, *37–50*, 37–51, *53–69*, 53–70, 122; Mack Avenue Plant, 34, *34*; Piquette Avenue Plant, 19, 20, 30, 35, *35*, 37, *37*; River Rouge Plant, 33, *72*, 72–73, 92
Martin, P. E., 93
Mass automobility, 122, 124–25
Mass production, *32*, 32–33, 36–37; at branch plants, 70–71, *70–71*; definition of, *122*; Ford's interest in, 16; at Highland Park Plant, 37–51, *38–50*, 53–61, *53–61*, 65–68, *65–68*; production output and cost related to, 50–51; of war material in World War II, 122
Materials: Ford manufacturing of, *69*, 69–70; steel, 19–20, 23, 31
Maxwell, 7, 17, *17*
Meaning of the Model T, 111–25; car of the century, 111, 121; as catalyst for change, 117; to collectors, 118–20; folklore about, 117–18; for hot rodding, 120–21, *121*; in jokes, *113*, 113–14; macro meaning, 121–25; mass automobility, 124–25; in music and songs, 114–15, *115*; naming of cars, 112–13; to owners, 117; in poetry, 115–16
Mercedes, 3–4, *4*
Mercedes-type gasoline cars, 3–4; advantages of, 12; cost of, 4; Thomas 1908, *6*
Metallurgists, 19, 21, 23
Michigan Socialist Party, 85
Model A, 14, *15*, 34
Model B, 14, *16*, 23, 24
Model C, 14, *15*, 34
Model F, 14, *15*
Model K, 14, 16, *16*, 20
Model N, 14, 16, 17, *17*, 18, 19, 20, 23, 24, 31, 35, 37
Model R, 18
Model S, 18
Model T: 1910, *5*, *11*; 1926 restyling of, 92–93; accessories for, 106–7, *106–7*; *Advance Catalog* for, 78, *79*; aftermarket equipment for, 107, *108–10*; as "car of the century," 111; collectors of, 118–20; colors of, 67; cost of, 12, 84, 91–92; creation of, 18–31; demand for, 78–79; driving of, 96–102; end of, 73, *73*, 93, 117, 125; falling behind the technological curve, 74, 86–87; Henry Ford's reluctance to give up on, 93–94; for hot rodding, 120–21, *121*; introduction of, 1, 12; lack of prestige, 87; life cycle of, 1, 74; maintenance of, 103–4, *104–5*; meaning of, 111–25; naming of, 112–13; operation of, 30; owner's manual for, 103; peak sales of, 73; as power source for farming, 107, *109–10*; price cuts for, 91–92, *92*; production of, 32–73; quantity produced, 73; registered number of, 117; restoration of, 118–20; selling of, 74–95; shipment of, 61, 62; speed parts for, 120; weight of, 21–22, 31, 39, 83, 87
*Model T Ford Car, The*, 23, 104
Model T Ford Club, 120
*Model T Times*, *119*, 120
Moog, Otto, 33
Motor assembly line, 54, *54*
Music and songs, 114–15, *115*
*My Life and Work*, 94

National Grange, 9
Nevins, Allan, 20, 34

New car financing, 84–85
*New Yorker, The,* 118
Night driving, 101–2

Oil level, checking of, 102–3
Oil prices and availability, 125
Opposition to automobile, *8,* 8–9
*Outing,* 12
Overland, 21, 84
Owner's manual, 103, *105*

Pagé, Victor, 104
Painting: of car bodies, 65, *65, 67, 67*; of wheels, 47, *47*
Piquette Avenue Plant, 19, 20, 30, 35, *35, 37, 37,* 50
Planetary transmission, *27,* 27–28, 86–87; view in Model T owner's manual, 103, *105*
Porsche 911, 111
Production of Model T, 32–73; design changes, 74; "disassembly line" principle, 52; duration of, 33; dynamic nature of, 33; effect of assembly line on production output and cost, 50–51; end of, 73, *73,* 93, 117, 125; at Highland Park Plant, *32,* 32–33, *37–50,* 37–51, 53–70, *63–69* (*See also* Highland Park Plant); quantity produced, 73
Prototypes for Model T, 20
Purchase of Model T, 83–84, 90

Quadricycle, *14,* 18, 19

R. L. Polk and Company, 117
Racing the Model T, 82–83
Radiator: assembling core of, 46, *46*; installation of, 60, *60*
Rebates to buyers, 83
Recreational driving, 10–12, *11*
Restoration of Model Ts, 118–20
Rivera, Diego, 33
River Rouge Plant, 33, *72,* 72–73, 92
Roads in America, 3–4, *5,* 23, 31, 83, 87, 124
*Roanoke News,* 117
Rockleman, Fred, 35
Rumely, Edward, 84, 85, 93
Runabouts, 3, 7; cost of, 7; Maxwell 1908, *7*

Savings plan for new car purchase, 90–91, *91*
Selden, George, 75
Selden patent suit, 75, *76*
Selling the Model T, 74–95; *Advance Catalog* for Model T, 78, *79*; advertisements, 79–81, *80, 82*; branches and agencies for, 74, 77–78, 86; on credit, 83–84, 90; demand for Model T, 78–79; *Ford Times,* 78, 80, 81; Ford Weekly Purchase Plan, 90–91, *91*; Hawkins' contributions to, 74, 76–78, 80, 82; by lowering price, 91–92, *92*; mass consumption and, 122; peak sales in 1923, 73, 92; profits on, 92; rebates to buyers, 83; terminal decline in sales, 92–93; track racing, 83; transcontinental auto race, 82–83; to women, 82, 87, 89–90, *90*
*Shape of Time, The,* 118
Shawmut, 83
Sheldrick, Laurence, 93
Shipping cars, 61, 62, 78
Side curtains, 102, *103*
Skill level of workers, 123
Smith, C. J., 20
Smith, J. Kent, 19
Songs and music, 114–15, *115*
Sorensen, Charles, 19, 20, 36, 37, 93
Spark lever, *97,* 98, 100
Spindletop oil strike, 125
Sprague Electric Works, 39
Stanley runabout, 3
Starting the Model T, 61, *61,* 96, *96,* 100–101; in cold weather, 101; electric self-starter, 101
Station assembly, 34; of Model N, 35, *35*; of Model T, 50, *50*
Steam cars, 2–3, 7; cost of, 2–3; disadvantages of, 12; White 1908, *3*
Stearns, 83
Steel, 31; vanadium, 19–20, 23
Steering, 30, *30,* 99
*Story of the Race, The,* 83
Strelow, Albert, 34
Strout, Richard L., 106, 111, 118
Style changes, 87–88, 94–95
Suburbs, 124

Textile department, *69*
Thomas, *6,* 22
Throttle, 97, *97,* 98, 99, 100
Tin Lizzie, 112, 113
Traffic fatalities, 125
Transcontinental auto race, 82–83
Transmission, 27; cover stamping, 45, *45*; planetary, *27,* 27–28, *29,* 86–87; sliding gear, 27, 28, 86; view in Model T owner's manual, 103, *105*

Unionization, 124
Upholstery installation, 65, 67
Urban transportation, 9
Used cars: expanding market for, 87; financing of, 84

Vanadium steel, 19–20, 23
Varnishing car bodies, 67
*Vintage Ford, 119,* 120
Volkswagen Beetle, 111

Wages. *See* Five-dollar day wage
*Wall Street Journal,* 85
Wandersee, John, 21, 23
Watts, Steven, 94
Wayne Automobile Company, 36
Weight of automobiles, 21–22, 31, 83, 87
"Western buggy," 4, 7; cost of, 4, 9; disadvantages of, 7; Holsman 1908, *6*
Westinghouse Airbrake Company, 41
Wheels: installation of, 60, *60*; painting of, 47, *47*

White, E. B., 106, 111, 118
White, Lee Strout (pseudonym). *See* Strout, Richard L.; White, E. B.
White steam cars, *3,* 40
Wills, C. Harold, 19, 20, 23, 74
Wilson, Woodrow, 9
Window/windshield glass manufacture, 69, 73
Windshield wipers, 102
Windsor, Canada, branch plant, 71, 72
Winter driving, 102
Winton Motor Carriage Company, 2
Wire mill, *69*
Wollering, Max, 36, 42
*Woman and the Ford, The,* 82
Women's market, 82, 87, 89–90, *90*